Glow
15

To all women →

from my daughter, Megan, to my mother, Valerie, for all my beautiful Latina women, and to those of you around the globe – thank you. Thank you for having the courage to be bold. Thank you for having the determination to not just survive, but thrive. Thank you for giving yourself the opportunity and strength to glow.

An Hachette UK Company
www.hachette.co.uk

First published in Great Britain in 2018 by Aster, a division of Octopus Publishing Group Ltd
Carmelite House, 50 Victoria Embankment, London EC4Y 0DZ
www.octopusbooks.co.uk

Published in the United States by Houghton Mifflin Harcourt Publishing Company, 3 Park Avenue, 19th Floor, New York, New York 10016

Published by special arrangement with Houghton Mifflin Harcourt Publishing Company

ISBN 978 1 91202 363 9

A CIP catalogue record for this book is available from the British Library.
Printed and bound in Great Britain

10 9 8 7 6 5 4 3 2 1

Book design by Melissa Lotfy

This book presents the research and ideas of its author. It is not intended to be a substitute for consultation with a professional healthcare practitioner. Consult with your healthcare practitioner before starting any diet or supplement regimen. The publisher and the author disclaim responsibility for any adverse effects resulting directly or indirectly from information contained in this book.

Glow 15

The 15-DAY programme based on the award-winning science of autophagy

NAOMI
WHITTEL

Contents

Foreword

I have spent my entire career focused not only on how to save lives, but also how to extend lives. I have travelled the world covering global health crises, while also learning lessons from the happiest and healthiest people on the planet. I have operated on brains in war zones, and I have seen cutting-edge technological developments in my own hospital back in the US that change the way we think about our health – and change the way we treat and prevent problems.

Along the way, my appreciation for the human body and its amazing capabilities has grown, as has my fascination with what makes it optimally function. One of the most intriguing processes in that regard is autophagy. It's a term you have likely never heard, but the scientific community has been studying this for decades – in fact, in 2016 a Japanese cell biologist, Yoshinori Ohsumi, won the Nobel Prize for his research on autophagy. Literally translated, it means self-devour, which sounds scary, but it is the body's natural way of recycling. When your autophagy is working properly, old and diseased cells are devoured or digested and then used to create energy and the building blocks for new cells. Dysfunctional autophagy, on the other hand, is linked to all sorts of age-related problems, including diabetes, cancer and dementia.

Like most people, when you read this you are probably already wondering how to tip the scales in your favour.

And, that is why I "devoured" my friend Naomi Whittel's book. It presents a commonsense, scientifically based programme to maximize your body's ability to repair its cells and fend off the damage associated with ageing. Whether it is an autophagy-activating tea, figuring out how to exercise smarter, or finally understanding your sleep type, there are simple, cheap and effective things we can do to increase autophagy and slow down the pace at which we age.

Naomi has been a leading innovator in the natural products industry. We first met when I began researching the merits of resveratrol, the compound found in grape skins and wine that has been linked to anti-ageing and disease prevention. Naomi and I then bonded over our conviction that deep within all of us is our ability to heal, if we just nurture the body to function normally. Armed with that belief, Naomi has travelled the globe in search of the purest and most potent ingredients, brought emerging scientific solutions mainstream and helped countless people not just survive, but thrive.

Naomi is one of a kind, a wellness warrior. Besides her incredible success in the natural products industry, she has a gift for making complex science understandable and relevant. Glow15 is a product of her work with leading doctors and researchers to create a programme that essentially hacks your biology to help you to look, feel and perform your best.

Sanjay Gupta, MD
Chief Medical Correspondent, CNN

Introduction

If you are one of the millions of women who are concerned about and experiencing accelerated ageing, you have good reason to feel anxious, but you are not alone. *Ageing* isn't just some word we use to describe the process of approaching the three-letter word: *old*. Ageing is a natural inevitability that *can* come with some unwanted side effects – ones that can actually start at the age of twenty-five. Worst of all? Most of the time, you don't even know it's developing. All of this ageing happens inside your cells – with no symptoms, no warning signs, no weird rashes or coughs, no reason to see a doctor. But this premature process, which bubbles inside you for about ten years before you even first notice that you're getting older, is happening. And it's aggressive.

By the time you turn thirty-five, cellular ageing is compounded by outside forces like processed foods, poor-quality sleep, technology overload, environmental pollutants, sun exposure, anxiety and more. The constant destruction is what accelerates ageing at both a cellular level and then at an observable one – in the form of weight gain, dull skin, and low energy.

And that's about when you think it: "I look older, I feel older – how did I *get* older?"

If the cellular damage continues, so does this cycle – for decades, as your body experiences both invisible and visible signs of ageing.

Research, however, also knows something else about the way our bodies work: About 70 per cent of longevity is related to lifestyle. Therefore, it is controllable. Therefore, you can do something about it. Therefore, you can stop accelerated ageing in its tracks.

The solution: Glow15.

Glow15 is a science-backed lifestyle programme to reenergize, cleanse and repair your cells, making your body younger no matter your actual chronological age. After following Glow15, you will lose pounds and inches, find the energy you once had, and make your skin more beautiful and radiant.

You will stand tall and be noticed for all the right reasons. You will choose clothes that accentuate your body, not hide it. You will spend your nights resting and restoring, not tossing and turning. Your to-do list will become a got-it-done list. Your relationships will be healthy and strong, because *you* will be healthy and strong, too. You will go-go-go because you want to, not because you have to. You will thrive, not just survive.

You. Will. Glow.

How will you do it?

I have developed Glow15 to be a new and unprecedented plan that's based on cutting-edge Nobel Prize-winning science as well as years of research from leading medical doctors, PhDs and practitioners. It revolves around the missing link in anti-ageing science, so that you can be better able to lose weight, boost your skin's radiance, get better sleep than ever, have endless energy and experience cellular renewal.

In my 20-year career as a founder and entrepreneur of nutritional wellness companies, and years of being a Certified Nutritional Consultant (CNC), I have guided millions of women to improve their lives by helping them slow ageing. Now I have developed a programme that includes a lifestyle plan and method for restoring your youth and improving your life. To get there, you will learn about:

- New and delicious foods and ways to eat that will change the way your body works – and looks
- The most efficient and most effective forms of exercise that will burn fat faster
- Customized sleep habits so that you snooze easier and wake up recharged and ready to go
- Pure, clean, and nutritional ingredients from around the world to help improve your body inside *and* outside

Glow15 is a lifelong approach to restoring and maintaining your youth and beauty, and it begins with a fifteen-day kick-start that sets you on the right path. In those fifteen days, you can expect to see tangible results –

results that will inspire you to make Glow15 a part of your daily life.

Here's what you can expect in these first two weeks:

Rapid Weight Loss: Lose up to 7 pounds and 4 inches from your waist, and reduce your dress size in just fifteen days.

Increased Energy: You will feel more alert and focused during the day. I felt much more energy when I started integrating the principles of Glow15 into my life – and so did the many women who have done so, too.

Fewer Wrinkles: Reduce your wrinkles up to 30 per cent. See fewer lines and brown spots, and get smoother, softer skin.

Better Health: Because of the little changes you make, you'll have fewer sick days – and you'll be better able to help your body prevent future diseases.

More Strength: This will happen quite literally as your body changes to give you beautiful lean muscle, but you'll also see another kind of strength – a boost of confidence.

I created Glow15 through a combination of methods.

First and foremost, I had access to the world-leading researchers and scientists studying the missing link in anti-ageing. I became their human guinea pig to fully explore why we're ageing too fast – and, more important, what we can do to reverse it. From there, I was able to develop Glow15 – an easy-to-follow lifestyle plan based on the complex science. This is the first of its kind.

Second, I scientifically tested Glow15 on women just like you and me. The results were incredible. When I saw the benefits to their bodies, their skin, and their minds, I knew this was a programme that had youth-restoring powers.

Finally – and perhaps most important – I got here because I'm just like you. As a wife, mother, and entrepreneur, I felt the effects of ageing – sometimes very profoundly. And those experiences became not only the foundation of Glow15, but also what inspired me to find answers.

Answers that changed my life. And, I hope, answers that will change yours.

In Pursuit of Purity: My Path to Glow15

I was born on an organic/biodynamic farm in Switzerland. My father is a chemist, and my mother is an artist. Looking back, I think their main job, though, was being 100 per cent dedicated to making sure that everything I ingested, put on my body or breathed was as clean as possible. That's how we ate: cheeses, berries, fruits, vegetables, nuts, a little meat, nothing processed. Our food wrappers came in the form of skins and shells, not plastic packs with red letters luring you to some sugary destination. I had my first "green juice" before I was two (I still drink it to this day, more than forty years later). My parents followed the principles of philosopher Rudolf Steiner, who believed that a biodynamic farm was its own ecosystem that served as the basis of optimal nutrition. They practiced the science of clean living, as they were very conscious of environmental toxins and pollutants. They wanted me to eat organic foods, drink clean water and breathe unpolluted air, so that was the life I lived. They believed in the purity of food and body in a philosophical way.

From the time I was a little girl, I experienced the pursuit of a pure and healthy body, because I suffered from eczema, an autoimmune disease that resulted in red, itchy, bleeding skin. I struggled every day at that young age, and my parents did everything they could to help me keep it under control.

The irony of it all was that my eczema made me look very unhealthy, even though we did everything we could to make me healthy. In many ways, that had a profound impact on me. I understood early on how outside appearances matter to the world. I understood that sometimes, no matter your best intentions, your body can have its own voice. Practically speaking, the eczema made me do things I didn't want to do.

I covered myself with clothing from head to toe. I wanted to hide. I never went swimming in the summer with my friends. I would never use any sunscreens or lotions because they would irritate and make my skin hurt that much more.

My body rebelled against me, and I felt every bit of that biological battle.

When I was fourteen, I had a big crush on a boy named Clive. I was convinced that Clive was going to ask me to the spring dance. Every day, whether it was warm or cold, I wore long sleeves and trousers to cover my dry, cracking and bleeding skin, so I could stay as hidden as possible. But understandably my mother wanted me to have the self-confidence to be comfortable in my skin, which also meant showing it to the world. She also taught me that I shouldn't be ashamed of how I looked, so one day that spring, I decided why not, I would put on a T-shirt and shorts.

Well, Clive got a glimpse of my red, swollen, bleeding arms and legs, and he never asked me to the dance. Instead, he took my best friend. My confidence was shattered.

I was able to control my eczema over time. But more important, I eventually learned that my own health and my own self-worth were more important than anything else (especially the affection of a teenage boy). Being devoted to health and wellness became the fabric of my life. My passion for health and wellness became who I was, what I believed, and how I felt. It made me feel good inside out and outside in.

I made it my life's work and career. I started in the health industry while still in college (even owning a health food store in my early twenties), and I eventually went on to develop nutritional supplements and to become an entrepreneur and founder of several wellness companies dedicated to helping women live more youthful and healthier lives.

But something happened. And that something is what directly led me to where I am today – and to the book you're reading right now. I remember the moment clearly: I was at my yearly visit with my integrative medical doctor, but this was a special visit because I was excited about getting my body ready to have a child. My doctor was excited for me as well. He did the full workup: bloodwork, urine testing, a full physical – everything. The visit was a formality but it was also going to be the go-ahead. When I got the call from the doctor's office, it wasn't to give me the results – it was a request to come back in. That was strange. I would normally receive a call that everything was great. When I went back in, I was feeling a bit nervous, but not expecting anything abnormal, because I did everything right.

When the doctor sat down with me, he told me that I had three heavy metals in my body – mercury, lead and cadmium. The levels were very high; in fact, they were off the charts. He was puzzled, and I was in shock.

My body was riddled with contaminants and toxins. I could not begin to understand why.

My doctor said that in no way did I have permission to get pregnant.

He started to go through a whole list of how this could have happened. He asked me if I had been exposed to heavy metals. No, not that I was aware of, I told him. To treat my eczema, I had tried prescription medicine, creams, pills, the whole gamut, but ultimately I worked with a Chinese physician and used acupuncture and Chinese herbs.

The herbal combination I had been prescribed included such ingredients as gardenia, mulberry leaf and chrysanthemum, which are harmless. I used them in baths and sipped them as tea, and within about eighteen months, my eczema had cleared up.

The lightbulb went off in my doctor's head – it was the herbs; they must have been contaminated. Together we realized that those ingredients may not have come from a clean source. They could have been exposed to pesticides or other environmental toxins while they were being grown or processed.

I was frustrated and felt foolish. I thought I had found a natural, healthy solution. I had tried to keep my body as pure as possible because I wanted to control my eczema. This news was unbelievable. I had paid so much attention to what I had put in my body, yet I fell victim to nutritional contamination. I knew right then how easy it was for my – or any woman's – good intentions to go wrong.

I made a promise to myself that day that I would always know where ingredients came from, and I would do what I could to make sure that everything I put in my body was pure. I wasn't going to be fooled again.

My doctor placed me on a detox plan, and in six months, my levels had started to return to normal. Today, I am a mother of four (yes, *four*) amazing kids. And I haven't forgotten my promise – to find the purest ingredients – and made it my mission to make those ingredients and products accessible to all women.

That dream was first realized after dinner with my cousin Dr Eric Lafforgue in Montpellier, France. That night, after I passed on a second glass of wine and another helping of rich, creamy cheese, Eric and his wife, Ude, helped themselves to seconds, then thirds. We joked about how indulgent this was; yet they, like most French people, were thin and healthy. This is the "French paradox", a term coined almost forty years ago around the idea that French people have a low rate of heart disease despite indulging in a high-saturated-fat diet. This was contrary to the belief that a major risk factor for coronary disease was a diet high in saturated fat – especially French indulgences like meat, butter and dairy.

A leading explanation according to scientists: resveratrol, an antioxidant found in red wine that has been shown to increase longevity. It was then that I decided to bring resveratrol to American women so that they, too, could benefit from its anti-ageing properties. I began Reserveage Nutrition, an anti-ageing resveratrol and beauty-from-within company, and continued to search for ways to make our lives healthier.

I travel the globe for new wellness- and youth-promoting ingredients. When I discover something new, I work with world-leading scientists to learn about the health benefits of the nutrient. I visit the source to see first-hand where and how it's grown and the level of purity as well as to establish sustainable partnerships. This helps me to ensure quality, safety and efficacy.

This practice of hands-on learning of where products come from is where things got *really* interesting.

The Missing Link in Anti-ageing

In my global search to find the best and purest ingredients to support wellness and beauty, I travelled to Calabria, Italy, to learn more about the powerful bergamot citrus fruit. I knew of its link to heart health, cholesterol and longevity, but I soon discovered its connection to a cellular cleansing process that plays a major role in ageing.

There, I met Dr Elzbieta Janda, a molecular biologist at Magna Graecia University. She explained that bergamot has a high concentration of flavonoids, which can help maintain healthy cholesterol and blood sugar. Then, she took a bite of the white portion of the rind, the pith – you know that part of a fruit you usually try to remove? – and said it was also a potent anti-ageing solution because it activated a process I had never heard of:

Autophagy.

Aw-ta-fo-*what?* As hard as I tried to pronounce it, I couldn't get it right at first. (It's *aw-TOFF-uh-gee,* by the way.) But I was fascinated.

Dr Janda explained that in addition to the natural wear and tear that our bodies experience, environmental toxins accumulate in our cells, accelerating the signs of ageing. Autophagy is the cellular cleansing process that removes those toxins and repairs the damage left behind. Dr Janda called it the fundamental process to prevent ageing because it helps keep your cells functioning at their best.

That first conversation led me to want to know more, to learn more and study more about the role autophagy plays in the ageing process. My exploration with the world's leading researchers, doctors, scientists and behavioural experts allowed me to appreciate the profound impact of autophagy. Like many things in your body, as you age, autophagy slows and becomes less efficient, making you look and feel older. My *Aha!* moment came when I learned that we don't have to accept that decline. We each have the power to reactivate our autophagy and make our cells behave younger again. This is why autophagy is the missing link in anti-ageing.

And it's also why I developed Glow15 – to help all of us defy the signs and symptoms of ageing. Glow15 is my answer to the question I get asked most often: How do you balance being a mother, being an entrepreneur, endless travel, maintaining your energy, and looking and feeling your best? This programme has unlocked my youth and my energy, allowing me to raise a family of four, travel the world, run a global nutritional organization, and appear as the resident nutritional expert on TV, along with discovering new nutritional ingredients.

Like so many of us, I have experienced the invisible, as well as the slow and visible, accumulation of signs of ageing. After I turned thirty-five, my weight started to creep up and I had moments when I could see the older version of myself in the mirror; there were times when all I wanted to do as I looked at the day ahead of me was crawl back into bed. I didn't have the energy.

But when I started doing things that activate autophagy – the things that are now the foundation of Glow15 – it changed the way my body worked. I made these small changes, and I ended up losing weight and gaining energy and my body fat percentage dropped. My skin glowed, and I *looked* younger.

Glow15 helps us hack our autophagy to reverse the signs of ageing, to make us look thinner and feel better, to give us the energy we need to perform at the top of our game, no matter what game that is.

We can stay younger for longer.

So what makes Glow15 different?

Throughout the book, you'll learn not only about the Nobel Prize-winning science behind the programme, but also how and why the programme works – and what you can do to influence how efficiently your cells work. And you'll also learn about some of the unique – yet easy – strategies that make up the foundation of the programme.

Glow15 takes into account every aspect of your lifestyle so you can seamlessly incorporate the plan into your busy life.

Diet: Discover how an innovative combination of intermittent fasting and protein cycling can help you lose weight and grow younger.

Nutrition: Learn about the powerful polyphenols that can help defend against and repair age-related damage in your cells.

Energy: Drink a specially formulated youth-activating tea packed with nutrient-dense polyphenols that will keep you energized throughout the day while boosting your metabolism and immunity.

Exercise: Try the most efficient, scientifically proven workouts for youth-maximizing benefits to keep your brain and body young.

Sleep: Find out why less may be more. Identify your sleep type to create a customized routine for your most restorative, restful nights yet.

Glow: Learn my secrets for getting radiant skin by using the most rejuvenating nutritional ingredients.

As you embark on Glow15, I want you to embrace both the science and the strategy. I want you to eat, drink and bathe yourself in the best natural resources. I want you to remember that what I am asking you to do won't require you to make too many changes in your life, although it will significantly change your life. I want you to open your mind to the possibilities, knowing that countless women have succeeded using this programme, and I believe that you will, too.

You will glow. You will radiate health. And you'll do it every day and in every way.

Chapter 1

Why Glow15?

The Scientifically Proven Plan to Look and Feel Younger in Just 15 Days

"You're never too old to become younger." — MAE WEST

You never forget your first time . . .

For me, it was shortly before I turned forty. I was rushing to get ready in the morning and had my usual 5 minutes to throw on some makeup. About 15 minutes later, my kids were late for school because I had to apply more of everything — more moisturizer, more concealer, more coverage. I had spent so much extra time trying to hide my lines and discolouration and working hard to brighten my dull skin.

That was my first time: The first time I realized I looked older. Felt older. Was I old?

For my friend Jenny, it was at thirty-eight, when a stranger called her "ma'am" instead of "miss." For my cousin Rachel, forty-three, it was when she found her first grey hair . . . in her eyebrow.

For my colleague Suzanne, thirty-four, it was when her husband asked her if she was tired in the morning after she had had a full, restful night's sleep. He said her dark circles suggested otherwise. It was the first time she realized she looked older.

The funny thing about it is that it's not about anyone else actually noticing; but when you experience your "first time," the way you feel about yourself changes.

Complain about it to your best friend, notice it in the rearview mirror, start pulling your face back contortionist-style to see what surgery could do? When this type of ageing starts, is there anything you can do to delay and reverse it?

The answer is yes.

And it's the reason I created Glow15 – to make your first time your last time. While getting older may be inevitable, you do have the power to change the way you age.

Glow15 Can Increase Your Youth Span

This revolutionary, science-backed programme is designed to increase the amount of time in years that you live with vibrancy, with energy, with focus and with glowing radiance. It will empower you to improve your health, your beauty and your confidence.

In just fifteen days, you can go from:

- unhealthy to healthy
- exhausted to energized
- restless to restful nights
- dull, damaged to supple, smooth skin
- foggy to focused

Glow15 is the culmination of scientific research and life-changing advice from world-leading medical doctors, PhDs, dermatologists, sleep doctors, nutritionists and fitness experts. It is the first plan of its kind that gives you the power to outsmart the ageing process and recapture your youth. Harnessing the power of autophagy, Glow15 keeps you looking and feeling young.

Autophagy boosts your body's innate ability to repair itself from within, allowing you to rid your cells of the damage that causes ageing; improve cell turnover for more efficient and effective function; and regulate your immune system to fight future infections, reduce inflammation, and help lower your risk for cancer, heart disease, and neurodegenerative diseases.

Methods of boosting this process have been studied for years at leading institutions and universities all over the world but have not been practiced outside academia or become part of the mainstream conversation – until now. Glow15, based on groundbreaking 2016 Nobel Prize-winning scientific research, is the first plan to pull the curtain back on this less-recognized but fundamental system that keeps your cells young.

Glow15 is a lifestyle plan specifically designed to work for your life – so that you do more than just survive, you thrive. Here's what I mean by that. In life, the green light is always on. We go. We go again. And then we keep going. The many complexities of our lives are beyond demanding. Our jobs require us to be focused. Our families – the heart and soul of who we are – rely on us to be energetic and affectionate. Our friends need us to be strong and supportive. And our drive to meet these demands can't be properly summed up in a few sentences, in 140 characters or in a funny Facebook photo.

We want it all. Too often, we do it all.

We have multiple jobs – whether it's at work or at home or both. We have lots of stress, arguably too much of it, and we get worn down by the constant buzz of life.

So what happens when we have all this pressure and all these responsibilities and all these dreams that we haven't quite got around to pursuing yet and our bodies – bombarded by pollution, environmental toxins and contaminants – start to look and feel older than they should?

We carry the weight of the world, and it's pretty darn heavy. It's damaging our cells, which can lead to weight gain, low energy, and dull and dry skin – the telltale signs of premature ageing. Life can become even harder to keep going, let alone going at the level that we want and deserve. That green light starts to dim.

But that doesn't have to be the case. Glow15 enables you not only to keep going, but to go better, to go stronger, to go smarter.

Putting Glow15 to the Test

I worked with an incredible team to take complicated science – which until now has mostly been practiced in research labs – and translate it into an easy-to-follow, user-friendly programme that works with our fast-paced lives.

Leading nutritionists and registered dietitians supported and developed a food plan with an understanding of how we really eat. Whether you usually prepare dinner for your family or spend most nights eating out for business, Glow15 can work for you. You will eat real food, available in real shops for real budgets. Plus, you will have real indulgences like chocolate and red wine. Though it's not designed as a weight-loss plan, Glow15 works to boost your autophagy to help you burn fat and change your shape.

I collaborated with a group of world-class researchers – medical doctors, PhDs, scientists and innovators – to navigate the exciting world of polyphenols. These nutrient-dense powerhouses boost your autophagy and work to reduce inflammation, increase immunity and fight the signs of accelerated ageing. Sourced from plants, they are readily available in vegetables, spices, teas, nuts, seeds and oils. We've identified the most powerful polyphenols to supplement your diet. They not only repair cell damage that can lead to disease, but also work to prevent it.

Fitness experts and exercise physiologists took abstract ideas and helped adapt them into moves you can do to look and feel younger. You won't need a gym or any high-tech equipment. No matter your fitness level, these routines are both easy to follow and appropriately challenging so you can get more out of your workout (and do less).

Renowned sleep experts and sleep doctors were consulted to create a customized sleep plan to help you determine your sleep type. By making the appropriate adjustments, you can drastically improve your sleep quality, allowing your body to repair and recharge overnight so you wake up more energized and focused, ready to tackle the day.

And skin experts – from dermatologists and leading chemists to aestheticians – assisted in assembling the ideal list of ingredients for rejuvenated skin. In addition, they helped to identify which ingredients lead to premature wrinkles and dullness and created DIY treatments guaranteed to give you back your glow. Collectively, we created a plan that we knew worked in theory; but more important, I wanted to test it and find out how well it worked in actuality. I was the first guinea pig.

I tested the plan. And I tested it again. And then I tested it some more!

I know Glow15 can work for you.

Because it worked for me.

As I mentioned earlier, I began to experience the signs of ageing in my mid-thirties. First it was a few extra pounds, then drier and duller skin, and a new feeling of blah – not really exhaustion, but no longer revving to go. And just as I began testing the Glow15 programme, I became the CEO of a publicly traded nutrition and wellness company. The pressure of the new job, coupled with the demands of my family, created more stress than I'd ever experienced. But instead of seeing increased signs of ageing – as should be expected with added anxiety – the opposite held true. The stubborn weight I had gained disappeared, my skin became clearer and brighter, and my energy levels tripled!

I was so thrilled with my results that my inner circle of family and friends wanted to try the programme. They, too, were blown away by the changes they saw in just fifteen days. My mother, Valerie, lost 8 pounds of fat and gained 2 pounds of lean muscle. Plus, she said she hadn't slept so well since before I was born! My sister-in-law Karen lost 4.5 pounds and her body fat percentage dropped from 28 per cent to 26 per cent. She was particularly impressed with how much smoother and more even her complexion looked. And one of my closest girlfriends, Monique, reduced her body fat by 10 per cent. While no one told her she looked tired before she started Glow15, Monique couldn't believe how many people commented on how well-rested she looked after fifteen days on the programme.

This was what the scientists and researchers had explained was possible with autophagy – to unlock the power of youth by reactivating your cellular process. And now that possibility became an actuality.

Clinically Tested, Woman Approved

I was excited with my inner circle's results. It was a promising start, but I wanted to prove Glow15 worked with women on a larger scale and in a controlled setting. My goal: to evaluate the plan and then improve upon it.

Heather Hausenblas, PhD, an internationally renowned expert in the link between physical activity and healthy ageing and the associate dean of the School of Applied Health Sciences at Jacksonville University (JU) in Florida, developed a clinical study in conjunction with the Brooks Rehabilitation College of Healthcare Sciences at JU to test Glow15 on thirty-four women between the ages of thirty-five and sixty.

They represented all of us – nurses, deans, stay-at-home mothers, CEOs, administrative assistants, lawyers – a diverse group of women with busy lives. They work, have families and are constantly faced with the increasing health struggles that come with ageing. The VISIA state-of-the-art facial imaging system was used to provide quantitative skin analysis. Changes in weight, assessing whole-body densitometry using air displacement via the BOD POD, were calculated in pounds, fat loss and body composition. Personal accounts and professional evaluations were collected. The data was recorded at fifteen, thirty and sixty days.

Dr Hausenblas was impressed by the findings. She concluded that the plan had profoundly positive changes on the participants' health. "I was surprised not only by their weight loss, skin improvements and increased energy, but by their enthusiasm to embrace the lifestyle changes. They seemed empowered by [the study] and motivated to continue practicing it even after it ended."

During the course of the study, 100 per cent of the women lost fat mass. Dr Hausenblas explained that changes were not only in pounds measured on the scale, but that participants actually gained muscle mass and improved their total body composition. More than 85 per cent saw a reduction in their body mass index (BMI). The women's body fat dropped by approximately 10 per cent after sixty days, and they lost an average of 9 pounds – some losing up to 7 pounds in the first fifteen days.

In addition, the women reported significant improvements in their skin firmness, complexion, youthful appearance, glow, pores, fine lines, elasticity, wrinkles, smoothness, crow's feet and tone. A dermatological assessment found an improvent in overall tone and a noticeable reduction in wrinkles. More than 90 per cent lowered their blood pressure – systolic BP improved from 127.31 to 122; and more than 80 per cent lowered their heart rate – on average from 74 to 70 beats per minute. Plus, a significant improvement in the women's sleep quality was noted. The average sleep quality rating changed from "fairly good" to "very good". In addition, participants reported positive mood changes, less anxiety and an increase in energy and overall health.

The numbers show that Glow15 scientifically and clinically works; but it's more important to know that it can work in *your* life. And I'm happy to report that the study participants found that to be true – like Lindsay, a working mother in her mid-thirties who had had three kids in the past four years. She told me that before she started the study, she couldn't remember the last time she did something for herself. "Between my kids, my job, my husband, and my friends, I felt guilty about spending any time on me. I'd only look in a mirror to put on makeup, and that was really for my coworkers – to spare them the sight of my dark circles and blotchy skin."

She admitted to being a bit overwhelmed for the first fifteen days and feeling guilty for spending time on herself by changing her diet, exercise and skincare routines. In fact, she confessed, "I mean, I'm a little embarrassed to admit how often I went to bed without washing my face. Now, well, my glow shows! I can actually leave the house without makeup and show my naked face. And when I choose to wear it – love that it's now a choice – my foundation goes on much more smoothly." Lindsay's fifteen-day imaging results revealed a nearly 30 per cent reduction in brown spots, 25 per cent reduction in pore size and overall improved colour.

Lindsay felt other changes before she saw them in the mirror. Her clothes got looser and her body fat dropped – going down two dress sizes in thirty days. During the course of the study, Lindsay continued to build lean muscle while her body fat percentage decreased by 10 per cent. Now Lindsay says the confidence she feels is more important than

GLOW15 SUCCESS STORY

"I finally put myself first" LINDSAY

"It wasn't until I started Glow15 that I realized how much I let myself go. Now **I can't believe how far I've come!**

"Between my kids, my job, my husband and my friends – I felt guilty about spending any time on myself. I'd only look in a mirror to put on makeup, and that was really for my coworkers – to spare them the sight of my dark circles and blotchy skin. The amazing thing is, now that I started taking care of myself, I feel like I'm so much more present and engaged with everyone else. Don't get me wrong, it was a lot of change at once for me, and that was a bit overwhelming – from the diet to the exercise to skincare. I mean, I'm a little embarrassed to admit how often I went to bed without washing my face. Now, well, my glow shows! I leave the house without makeup, and when I choose to wear it (I love that it's now a choice), my foundation goes on much more smoothly. **With Glow15, I've seen a 30 per cent reduction in my brown spots, a 25 per cent reduction in my pore size, and overall improved colour and fewer wrinkles in just 15 days! I've gone down two dress sizes** and the confidence I feel is more important than any number on the inside of my jeans."

any number on the inside of her jeans. "Honestly, it wasn't until I started Glow15 that I realized just how much I had let myself go," she says, "but now I can't believe how far I've come." What I love about Lindsay's story is how it echoes the study's larger-scale results.

The researchers at JU were particularly inspired by the compliance of the women. Of the thirty-four women, thirty-three completed the programme, and the one woman who had to drop out after thirty days, because of an unrelated health issue, asked if she could continue on her own. Dr Hausenblas said, "Adherence rates like this are very rare, especially with behavioural intervention, where there are so many different changes. People really liked the programme and noticed results quickly, so they stayed." That showed the researchers and me that the plan was something that was both doable, enjoyable and effective.

The women in the study benefited from Glow15 because of its power to work naturally in harmony with your body – boosting autophagy to enhance vitality. I wanted – and always want – women to experience the results of healthy weight loss, more lean muscle mass, improved skin function and improvement in markers that show risk of disease. So I've continued to share the programme (and continued to test and tweak it) with women all over the country. You'll see some of them reflected in the pages of this book. Shawn, a forty-six-year-old mother of four and business owner, told me she had tried everything to lose weight – different diets, different exercises, different everything. And it all failed. She kept trying to hide her body, wearing big shirts "always untucked" because she didn't want anyone to see her bulges. "My control-top tights tightened my breath and circulation more than my tummy," she said.

But once Shawn tried Glow15, her transformation began. She took 4 inches off her waist, going from 33 inches to 29. "I'm tucking my shirts in now!" And she lost more than 7 pounds in the first 15 days. She even lost 13 per cent of her body fat – dropping from nearly 33 per cent to 28.5 per cent – all in just 15 days. Those changes created momentum for her to keep going.

Shawn recognized that some parts of the programme were an adjustment, but now she can't imagine life without it. "Glow15 is no longer just a plan for me," she says. "It is my lifestyle."

Stories like Shawn's inspire me, their results drive me, and my hope is that they will motivate you to glow, too. By following a lifestyle that activates autophagy, countless women have seen immediate results – in all aspects of their lives.

If you answer yes to more than one of the following questions, I believe Glow15 can help you, too.

GLOW15 SUCCESS STORY

"No more love handles" SHAWN

"For me, **Glow15 succeeded where everything else failed.** As long as I can remember, I've fought the fat around my stomach – and no matter what I tried . . . foods, exercises, wishing and hoping (!) I kept losing the battle. Before Glow15, I had not so endearingly referred to the flesh hanging over my jeans as 'love handles.' I wore big shirts and never tucked them into my trousers for fear someone would see the bulges. My control-top tights tightened my breath and circulation more than my tummy, but I depended on them too. Not anymore – while I need to go shopping for smaller size clothing, my new hip bones (who knew they were under there?!) are holding my trousers up! I'm not going to pretend this was an easy overnight change – intermittent fasting made me 'hangry' and protein cycling took some getting used to – but **I kept at it and lost over 7 pounds in 15 days.** More than that, **I took 4 inches off my waist (yes, I'm tucking my shirts in now!), lost over 13 per cent of my body fat, and decreased my body mass by 2 per cent.** Glow15 is no longer just a plan for me, it is my lifestyle. Every Sunday I prepare for the upcoming week. I use the shopping list for activating foods and try putting my spin on recipes. Plus, I always fill two muffin tins with an Egg15 recipe (so far, my favourite is adding cheese on High days and peppers and onions on Low days) – and I love that those are now the only 'muffin tops' in my home!"

1. Have you put on weight in recent years that you can't lose?
2. Do you think of food often or feel like you are constantly dieting?
3. Do you have elevated blood sugar, blood pressure or cholesterol levels?

4. Do you have a family history of cancer, heart disease, diabetes, or neurological conditions like multiple sclerosis, Huntington's disease or Parkinson's disease?

5. Do you crave more energy and stamina in your day?

6. Is it hard to recover from exercise? Do you feel tired and sore the next day?

7. Is it hard to fall asleep or stay asleep?

8. Do you wake up more than twice per night to use the bathroom?

9. Does your skin look dull and your hair feel dry and limp?

10. Do you feel you need to be wearing makeup to leave your house?

Glow15 will help you defy not only the visible signs of ageing, but the invisible ones, too. You will boost your autophagy to combat the environmental toxins, outside stressors and other factors contributing to ageing.

By showing you the right foods to eat, the best nutritional supplements to take, the smartest ways to exercise (without spending a lot of time), the unique ways to improve your sleep, and the groundbreaking strategies for radiant skin, Glow15 will help you look and feel younger than you chronologically are.

Best of all? It takes only fifteen days to see significant changes.

Let's get glowing.

First, I want to help you understand more about the most important breakthrough in the science of ageing – autophagy.

Glow15 Results

At some point during the sixty-day Jacksonville University study,

- 87.5% of the women had an improvement in their **BMI**
- 87.5% of the women **lost weight**
- 96.9% of the women **lost body fat**
- 100% of the women **lost fat mass**
- 87.5% of the women had an **increase in fat-free mass**
- 90.3% of the women had an **improvement in their blood pressure**
- 81.3% of the women had an **improvement (decrease) in their heart rate**
- 90.3% of the women had **improved body satisfaction**
- 93.8% of the women had **improved skin satisfaction**
- 90.6% of the women had **improved sleep quality**
- 100% of the women had a **reduction in wrinkles**
- 91.3% of the women reported **smoother-feeling skin**
- 91% of the women reported **softer-feeling skin**

Chapter 2

What Is Autophagy?

The Most Important New Breakthrough in the Science of Ageing

On even the best of days, you stress about your family, your work, your friends and all your responsibilities. Now, picture a day when everyone and everything is even more demanding: You feel absolutely overwhelmed. Everything is coming down on you. Your boss gives you a new project with a ridiculous deadline, your kids are sick, your best friend is having a personal crisis, your computer crashes, your phone has no signal and someone just rear-ended you in the car park. Now let's pretend you haven't showered. Or slept. There is no relief at the end of the day or overnight. Your body and mind feel worn down and destroyed. What happens if the next day is exactly like this one? And then the next. And then the next. And so on.

At a cellular level, this is what ageing is: Your cells get beaten up by the stresses of daily living. But the overbearing bosses and fender-benders come in the form of such things as environmental toxins, processed foods and chronic stress. Over time, they wear on your cells, causing damage. That damage ultimately results in signs of ageing like weight gain, wrinkles, pigmentation, and even neurological diseases and cancer.

Now picture that same overwhelming day. At the end of it, instead of no sleep, no shower and nothing to make it better, you get the exact opposite: a nice warm bath, a glass of wine and restful, uninterrupted beauty sleep that leaves you revived and refreshed when you wake the next day. You feel like the you of your youth – alive, happy, ready for whatever awaits.

At a cellular level, *that* is autophagy. It cleans up the mess left behind from your overwhelming day and repairs any damage, restoring and rejuvenating your cells. Your cells function better, your body runs better, you look and feel better. Autophagy, a little known but fundamental process, is the key to anti-ageing and the foundation of Glow15. While you can't completely avoid the causes of ageing, you do have control over *how* your body ages. Glow15 will give you that power – and in this chapter I'll explain why boosting autophagy can help you defy ageing. My hope is that you will be as excited about the science behind Glow15 as I am to share it.

The journey that has led me to this moment reminds me of when I moved to the United States from England at the age of eleven. I was obsessed with braiding friendship bracelets like I had done for the past couple of years in the UK. Nobody in my new school had ever seen them. I knew my new classmates would love them. I started a school-wide trend, which helped me make friends quickly, and I realized how much I enjoyed sharing new things with others.

Thirty-plus years later, it's still thrilling to me. I love being able to pass along something new, often from a different part of the world, that can be translated into better health, wellness, beauty and strength. For my career, that sharing has come in the form of nutritional supplements that help improve women's health and well-being. In all, I have spent the last twenty years developing patented nutritional ingredients as well as science-based products designed to improve our natural beauty, increase our longevity, and restore energy and wellness.

Glow15 is my new friendship bracelet – but instead of wearing it on your wrist, you will be empowered by its ability to help you radiate better health from the inside out.

How We Age

What happens with an old phone? Things get a little slower, the apps don't work quite as well, sometimes they crash because they can't keep up, and we get so frustrated that we just wind up trashing it and upgrading to the newer, smarter, faster technological toy. In that scenario, the device experienced wear and tear, and just couldn't keep up with today's demands.

That sounds a lot like our bodies. As we get older, we want to keep doing what we were doing when we were in our twenties, but our insides – all our anatomical hardware and software – just can't keep up. Oh sure, it would be nice to upgrade to a new model every time things got a little slower ("Hey, time for the iBody X!"). And yes, when our screen cracks (in the form of new wrinkles!), it would be nice to go in and just replace it with a shiny new one. But our bodies are much more complex than even the smartest of phones. Our bodies can't be replaced. We need the ones we have to keep running as smoothly, quickly, and efficiently as possible, for as long as possible.

The question then becomes, if we can't trade in our bodies, how can we super-boost our insides and outsides – or upgrade them to not only last longer, but act newer?

To best understand how you can use autophagy to upgrade your body, let's look at why our bodies get slower and less efficient.

Ageing, of course, is a complex topic. There's not one thing that makes us go from young to old. An interwoven system of biological reactions can contribute to the way we age. We don't even see the deterioration happening at first – these cellular changes are invisible signs of ageing. We start slowing down before we, well, start slowing down. While we tend to think of ageing as a visible process – as we see wrinkles or weight gain – the reality is that ageing has already started at the cellular level. How?

There are a multitude of mechanisms that contribute to ageing, but I like to categorize them into two main groups: the Inevitable Agers and the Accelerated Agers.

The Inevitable Agers: These are, exactly as their name implies, the inevitable causes of ageing – the natural wear and tear on your body. The

Inevitable Agers reinforce the idea that over time, your body naturally starts to slow down and wear down. In the absence of some sort of intervention to stop it, our minds, our metabolism, our muscles, our joints, our eyesight, our hearing, and just about every other system in the body naturally slows.

The irony, of course, is that the human body is complex. Some parts of our bodies, if they weren't used at all, would wither up. It's the old "use it or lose it" mantra. That's the case with our brains; we need to constantly flex our mental muscles to keep them sharp. That's how memory is strengthened; when neurons are used to learn something new or perform mental tasks, the connections in the brain are strengthened. That's the case with our bones; in order to keep them strong, we need to put them through stress to regenerate new bone. That's the case for our lean muscle mass as well; muscle grows and fortifies when we use it.

The option isn't to *not* put our bodies through wear and tear. The very fact that you're living a full and active life means many of your cells and systems will degenerate *because* you're using them.

The Accelerated Agers: These are outside forces that damage your cells and speed up both the visible and invisible signs of ageing. While the Inevitable Agers cause the ageing-related degeneration you can't avoid, the Accelerated Agers are the exact opposite – you can avoid these factors that will wreak havoc on your body, leading not only to more wrinkles and weight gain, but also to age-related diseases like cancer, Alzheimer's and Parkinson's.

The most common Accelerated Agers are:

Foods with added sugar. Added sugar increases your body's demand for insulin – a hormone that helps your body convert food into energy. That excess insulin stresses your system, causing many different types of cellular malfunction, which leads to ageing-related diseases and conditions like diabetes, heart problems, and high blood pressure. Added sugar also reacts with protein, creating advanced glycation end-products (AGEs) that age the skin, causing skin to wrinkle and cell structures to harden. Think of the way arteries harden to cause coronary heart disease: A similar process takes place in the cells of the skin.

Robert H. Lustig, a professor of pediatrics in the Division of

Endocrinology at the University of California, San Francisco, is considered the leading authority on the dangers of sugar. Dr Lustig says that your body can safely metabolize up to 6 teaspoons of added sugar per day, yet the average American consumes 20.5 teaspoons (82 grams) every day. That translates to about 30kg (66 pounds) of added sugar per person, per year. It's not surprising that there's a chronic disease crisis.

Environmental toxins, such as excess UV exposure, pollution, and cigarette smoke. These toxins enter your body – via your mouth or skin – and damage your cells. Excessive sun exposure breaks down the cells to not only cause the sunburn you see on the outside, but also damage skin-related cells and structures to make the cellular environment toxic, leading to the development of skin cancer.

Other accelerators, including a sedentary lifestyle – sitting is the new smoking! – and lack of sleep, contribute to signs of ageing. According to the Sleep Health Foundation, nearly half of us aren't getting enough shut-eye, and women have a harder time sleeping and are more likely to feel unrested than men.

There's more to the story of ageing: Scientific research pinpoints the causes more specifically and has classified a number of different factors, systems, and mechanisms that contribute to the deterioration of our cells. They include inflammation, DNA damage, genetics, dysfunction of the mitochondria (the power plants of the cells), hormonal changes, neural degeneration, and muscle composition. The complex part about ageing is that there's no single cause – it's the interconnections among all these factors. It's not like you can just get a prescription to bulk up your mitochondria and, *Voilà!* – all is well.

But here's the good news: Glow15 is an approach that includes diet, exercise, sleep and skincare solutions (not just one magic bullet) to address the complexity of factors, causes, and relationships in the body that cause ageing.

Remember, you can control 70 per cent of ageing-related symptoms and problems by controlling the Accelerated Agers and staving off or slowing down the Inevitable Agers.

The Big Picture: The Origins of Autophagy

Battling the visible and invisible signs of ageing is very much a part of what led me on my search to better understand the science of autophagy.

In the 1950s, Belgian scientist Christian de Duve was studying insulin when he accidentally discovered a process he called autophagy, from the Greek words for "self" (*auto*) and "eating" (*phagy*). It is the mechanism by which cells cannibalize some of their own parts in a continual cleanup process.

In the 1970s and '80s, researchers began looking at the process of autophagy. It had not been studied extensively at the time, and nobody really knew its role or why it was important. The big breakthrough came in 1983, when researcher Yoshinori Ohsumi, while conducting experiments in yeast, discovered the genes that regulate autophagy. He found that without those genes, autophagy doesn't work – and the cells can't repair themselves. He won the Nobel Prize in 2016, as his work was considered fundamentally important to understanding how autophagy functions in cells.

The most fascinating part of the discovery of autophagy is that the process is given a boost if there is cellular stress. If cells lack nutrients, are deprived of energy, or are damaged in some way, a "stress response" mechanism is activated, which initiates autophagy. As a result, cell function actually improves when we're under stress. In the absence of added stress, autophagy remains functioning at a moderate level, maintaining cell function. This is known as its maintenance mode.

As the research evolves, more and more scientists will be looking at the role of autophagy in ageing – and how it is influenced by cellular stress, and that includes the Inevitable and Accelerated Agers.

On my quest to learn from the leading experts, I met William A. Dunn Jr., PhD, a professor of anatomy and cell biology at the University of Florida and one of the most respected authorities on autophagy, who has been studying the process for more than thirty years. He explained that as your cells get older, parts of the cells age and become nonfunctional. "Autophagy is a way of rejuvenating the cell," he said. "It basically gets rid of the nonfunctional components of the cell."

The thinking goes that when you activate autophagy, you reduce the chance of developing age-related problems and thus extend your life span. One study from the *Journal of Clinical Investigation* supports this theory by showing that things that extend longevity also show an increase in autophagy. That's partly because, as another study from the journal *Cell* shows, one of the characteristics of ageing is the accumulation of various forms of "molecular damage." The study authors noted that one of the most promising areas in combating that damage is autophagy.

How does it combat that damage? It's your cellular cleansing cleanup crew.

Autophagy 101: Your Cellular Cleansing Cleanup Crew

In most aspects of our life, we have no self-cleaners. Sure, there's the self-cleaning oven, and our computers can do it, too, with automatic virus scans and whatnot. But most other things require some time and attention to remove the grime, dirt and waste that accumulate over time. Wouldn't it be nice to have self-cleaning bathrooms? But think of what would happen if we didn't maintain our "things" with regular cleaning. Our kitchens, cars and desks would become toxic cesspools of dirt, and they would be sickening. That, in a way, is what can happen to our cells. If waste and toxins are not properly disposed of, our cell functions degrade and decline, resulting in everything from dull skin to lackluster energy to weight gain to age-related diseases.

This is exactly what autophagy, your cellular self-cleaner, addresses. Remember, it literally means "self-eating" because your cells eat away their own junk.

The result? Cleaner, younger, healthier cells.

Let's take a look at how it works more specifically: Our bodies are like little universes. But instead of stars, they're made up of trillions of cells, all of which play a role in how we function and how we live. They're made up of a variety of parts that influence those cellular functions. For example, we have the mitochondria, which generate energy for the cell. Cells also include proteins, which are essential for virtually all cellular functions: They give structure to cells, they carry out chemical reactions in the body, and they serve as messengers to communicate a variety of information across the body.

Though they're microscopic units, each cell chugs and churns, producing power and products that make your body function the way it does. These cells are doing the work that makes us think, move, feel – everything. These cells give you the power to send texts, remember song lyrics, calculate mortgage payments, write life-changing memos, reason logically with toddlers (well, maybe our systems haven't quite been able to master that yet!), and everything else you do throughout your days.

These cells are always working. Most of the time, they do great work, especially when they're young. Everything is brand-new, all systems are working in harmony, and things keep plugging along without a hitch. The cells do their jobs well and efficiently, and the end result is a young, energized, healthy body.

That doesn't mean each cellular system works perfectly all the time. Our normal cellular machinery gets damaged with use over time through those Inevitable and Accelerated Agers I explained earlier. Many people assume that wear and tear is a fact of life – that no matter what we do, our bodies are going to break down because of ageing. Certainly, we can't defy the natural arc of life and death, but we can indeed stall the effects of ageing. Here's why.

Your cells break down parts of themselves by sequestering them into vacuoles and digesting them. As a result, they produce waste, mainly dead organelles, damaged proteins, and oxidized particles, and that waste needs to be removed. But unless it's properly disposed of, that cellular waste stays and builds up in the body, becoming toxic to our cells. That accumulation is a key factor in the rate of ageing. The junk gets in the way and makes everything malfunction. It may sound like jargon from biology class, but the reality is that when the toxins damage the machinery of your cells, it contributes to the ageing process – it causes your skin to look older, your body to slow, your energy to drop and your hormones to go haywire.

I don't want that, and I know you don't either.

That's why autophagy is so important. It's like a cellular waste disposal, taking the dysfunctional parts and obliterating them so they don't cause a mess anywhere else. When it's working well, it's a form of self-renewal, breaking down older structures so that new ones can be built in their place. The result is that the newer, more youthful structures allow our cells to function as if they're, well, newer and more youthful!

You can imagine how that plays out in your everyday life. More youthful cells means softer and healthier skin, less fatigue, a faster metabolism – it means your universe of cells radiates youthfulness everywhere, from your organs to your brain to your muscles to your mental well-being.

I met with leading autophagy researcher Beth Levine, MD, director of the Center for Autophagy Research at the University of Texas Southwestern Medical Center. The expectation of her research is that it will lead to major discoveries in the treatment of ageing, diseases like cancer, and infectious diseases. Her lab identified the first known gene responsible for autophagy in mammals. She says the waste-removal function is what keeps us healthy.

In a study in the journal *Diabetes & Metabolism,* the authors described autophagy as a "survival strategy", which makes a lot of sense: from a cellular perspective, we're just trying to survive. From a lifestyle perspective, the strategy to survive becomes a strategy to thrive – because younger cells mean a younger self.

The Autophagy On/Off Switch

So why doesn't autophagy work well all the time? Autophagy is like every-thing else in your body: as you age, it naturally declines and becomes less efficient – meaning that you're now saddled with a twofold problem. Your cells accumulate a lot of junk and your body can't keep up with clearing away that junk due to your constant exposure to Accelerated Agers (see page 32). More waste equals more damage. More damage, with no way to fix it, equals increased ageing.

All that cellular waste is linked to not only things like skin damage, but also cancer and neurodegenerative diseases like Parkinson's. This decline in autophagy is considered a vital consequence of ageing. It's not just the deterioration of our cells that's so bad; it's also the decline in our ability to fix it.

There's another important point to consider: Autophagy can't always be on high. Cells can't be constantly cleaning up waste; they also have to be doing the work that *produces* the waste. Think about it like this: picture your kitchen. Let's say you make dinner and clean up after the meal. This multistep process includes preparing the meal and getting rid of the waste after you eat.

So autophagy is a series of steps that keep your cellular kitchen free of clutter, and you need them all to work. If there's a pileup at the sink or the kitchen bin or the outside dustbin, there will be clutter *somewhere*.

But the catch is that you can't always be in cleaning mode; you have to have something to clean up. You need the kitchen to prepare your food, just in the way that your cells need to be doing other jobs. So when a kitchen is truly humming, it's a balance between making food and cleaning waste. Ideally, that's the same process for your cells – providing energy to your whole body and then cleaning up the waste products. This is the idea behind autophagy being turned on and off. When the kitchen is being cleaned, in response to that stressor, autophagy is working at its highest level – remember the stress-response mode. And when you're making and eating dinner, autophagy is at its lowest level – the maintenance mode. Think of it this way – it's like electricity. If you plug in a lamp, the electricity is always flowing whether the switch is turned on or off, but more electricity is needed to power it on.

Glow15 will show you the best ways to turn autophagy on and off to defy ageing. There are two ways to do this: First, naturally induce some stress on your body; and second, integrate autophagy-activating nutrients. Since most of us, as a result of those Inevitable and Accelerated Agers, have low levels of autophagy, I'll show you how to boost your cellular cleansing cleanup crew's function so you look and feel younger.

Advanced Autophagy: The Missing Link in Anti-ageing

Autophagy is one of the most important breakthroughs in the science of ageing. While scientists have known about this process since the 1950s, only in the past decade have we been able to see the effects of activating autophagy to improve our cellular health. Research published in the *Journal of Clinical and Experimental Pathology* explains that "autophagy promotes cell maintenance by removing accumulated toxic material and by using recycled components as an alternative nutrient resource. This suggests that autophagy favours longevity because an organism can recover more quickly from stress-induced cellular damage." As you can imagine, when the cells can effectively clean up the damage that they produce, that means they will function better.

In practical terms, that means all of your body's cells will work better.

Autophagy enhances metabolic efficiency. From the deepest cellular level, autophagy can be activated to help improve the work of the mitochondria – the cell's power plant. This makes cells work more efficiently. By doing so, autophagy helps cells become more resilient.

Autophagy prevents neurodegenerative disorders. Many neurodegenerative disorders are the result of damaged proteins that form in and around neurons. Autophagy protects us by getting rid of these proteins. In Huntington's disease, Parkinson's, and Alzheimer's, autophagy cleans up specific proteins associated with those diseases.

Autophagy helps fight against infectious diseases. It does this by removing toxins that create infection, as well as by helping to improve how your body's immune system responds to infections. Intracellular bacteria and viruses can be removed by autophagy.

Autophagy improves muscle performance. When exercising, we place a stress on our cells, energy goes up and parts get worn out faster. Because of this, autophagy helps remove some of the damage and keep our energy needs in check.

Autophagy aids in the prevention of cancer growth. Autophagy suppresses systems and processes that can be related to the development of cancer, such as chronic inflammation and damaged DNA. Mice altered to have inefficient autophagy were found to have higher cancer rates.

However, cancer cells can hijack the process of autophagy and use it to their advantage to become chemo resistant. In later stages of cancer, as tumors begin to take over cells, autophagy is induced due to the stress of nutrient deprivation and energy depletion. This is why it is important to understand that autophagy should be turned on and off. Further research is needed since currently there is no way to measure the level of autophagy in humans when it comes to this area.

What science can tell us today is that autophagy works to make your body work better. By cleaning up cellular junk, you will clear the way for cells to rebuild themselves with new parts. Sort of like that biological upgrade – giving an older car a newer engine, so that it not only keeps running, it "corners like it's on rails" (that's a good thing, for those of you who haven't seen the movie *Pretty Woman*!).

How to Activate Autophagy: The Foundation of Glow15

In today's world, we tend to look at things as black or white – no grey. Things are either good or bad. But when you really think about it, life is full of examples that are both good *and* bad at the same time – even things that have positive effects can have negative ones. Take the smartphone. It has unbelievable powers, and the technology has made our lives so much easier, but some would also argue that it has eroded social skills and created many more burdens on families as well. Or about how the sun? It's unquestionably good because of its life-giving properties, but too much of it can kill us. What about fire? It provides warmth, it kills germs, and it can be essential for survival. But when left uncontrolled, fire can wreak havoc.

This "grey area" is essential to understanding how autophagy is activated via cellular stress.

Remember, autophagy is turned on when your body is in stress-response mode. Stressing your cells is the natural way to turn on autophagy. On Glow15, you'll do this through intermittent fasting and protein cycling (more on page 56) to create nutrient deprivation in your cells. You can also create cellular stress in the body through exercise (more on *that* on page 89).

I will show you novel ways to manufacture or hack your autophagy activation. By taking polyphenol-rich supplements and using unique nutritional ingredients, you can control your autophagy.

And that means you can influence the way you age.

GLOW15 SUCCESS STORY

"The first plan that's worked for me" KERRY

"Glow15 changed my life – and my body! **I lost almost 10 pounds in the first 15 days, and by the next 15 days my body fat percentage went down 3 per cent and my waist was 3 inches smaller.** Until my late thirties, I didn't really worry about my weight – but two kids, and too much eating off their plates later, the pounds just started creeping up. I got to a point where I almost didn't recognize myself. I tried a few things, but I didn't see the results fast enough, so I gave up. And then **I tried Glow15 – and the changes were almost immediate.** Everything about this plan was new for me. First of all, it worked – that was new (ha). A lot of the autophagy-activating foods were new: I'm 46 years old and I had never had an avocado – I didn't even know how to cut it. And quinoa was new too – I still can't pronounce it correctly and forget about spelling it – but I love to cook it. I didn't actually believe the numbers on my scales – they went down so quickly. People at work started asking what I was doing differently and telling me how great I looked. Now they have to listen to me 'fake complain' about my clothes being too big! Even my bra is bigger – but in a good way, I'm perkier now!"

Chapter 3

Foundations of Youth

The 5 Principles of Glow15

"What you can do, or dream you can, begin it. Boldness has genius, power, and magic in it !!!" — GOETHE

Glow15 is about being bold.

Yes, in so many ways, you're already bold — doing it all by balancing work and life, wearing the multiple hats of caretaker, teacher, employee, confidante, friend, parent, mentor. But I want you to believe that you can be even bolder.

Glow15 will help you do that. This plan is about you. And its principles are designed to help you not only survive, but thrive.

How do you get there? Above all, the boldest thing you can do is to *make yourself a priority*. While we've been conditioned to think that being bold is about aggressiveness or stomping over other people to get what we want, this is not what being bold is. My friend Lisa Schiffman, who helps women thrive by teaching them to scale and grow their businesses as the director of brand, marketing and communications at Ernst & Young, is passionate about empowering women to be bold. She says, "It means having the heart or the courage or the conviction to pursue a dream, stand up for something important, carve a path. Sometimes we hold back on bold out of uncertainty or concern for how we'll be perceived. But the most worthwhile challenges are often the ones that are hard."

I want you to remember this as you begin your journey on Glow15: Give yourself permission to be bold.

Be bold with courage. You've already shown the courage it takes to make changes by the mere fact that you're reading this and feel ready to start. That does, in fact, take courage – courage to attempt something you've never done and courage to do it even if it means you may feel temporarily uncomfortable. Yes, it will take courage to achieve your goals, but you gain courage with every new attempt. That allows you to be more courageous in your everyday life, which in turn will help you achieve your goals and aspirations.

Be bold with determination. You can pick whatever "one step at a time" cliché you want, but that's how life works. You don't achieve your goals by wishing yourself from point A to point Z in a blink. You get there by taking one step, then another, then another. And when you put forth that "I can" attitude, you will end up with an "I did" result. You absolutely have the power to make whatever changes you want in your life. Just take the first step, and the next one will follow.

Be bold by example. Throughout this book, you will read the stories of women who were bold enough to say "enough" to putting everyone else ahead of themselves. They decided it was "me time", and they got results, because they made a conscious decision to stop feeling guilty about taking care of themselves. Please let their examples lead you – and as you enjoy your own success, you will boldly lead others as well. We have an incredible power to lift each other up – with encouragement, with inspiration, with messages of determination. Sometimes those messages can be direct, and sometimes they can be subtler: a hug, an encouraging conversation, the right emoji at just the right time. Don't underestimate your power to not only help yourself, but to help others as well.

In the next few chapters, I'm going to take you through the major elements of Glow15 so you can get a full look at how diet, nutritional supplements, exercise, sleep and beauty work together to create the components of the programme. But in this chapter, I want to pull the curtain back on the core principles of Glow15. These guiding themes are at the heart of who I am and how I live – and I hope they will guide you, as well.

Principle 1: Outer Beauty Reflects Inner Health

In most areas of life, we *do* judge a book by its cover. The way something appears on the outside gives us a taste of what it's probably like on the inside. That goes for the major "things" in our lives, like cars and homes. That goes for actual books – you likely picked this one up because something about the cover resonated with you. It also goes for people. Yes, our human book covers include several factors, like our fashion and our general joie de vivre. Our appearance is tied to so many other factors in life. For example, it plays a role in first impressions, in confidence, in self-esteem and in our sexuality. All those roles have an effect on our health. Any negative emotions tied to your appearance can lead to very real physical issues. For example, being unhappy about how you look can make you anxious. That anxiety can lead to a suppressed immune system, extreme fatigue, high blood pressure, and stomach and back pain.

Your outer beauty reflects your inner health.

Many people assume that any talk of beauty is simply some superficial banter about creams, makeup and cover-up, lipstick and lip gloss. They assume that a woman's quest to be beautiful is about pleasing others. They assume that the only value in trying to look good is attraction. And they assume that any talk about skin is, well, just skin-deep.

Those assumptions, however, simply aren't true.

While beauty can certainly be about attraction and self-confidence, beauty *is* about health and wellness. The health of our skin is also a reflection of how healthy we are overall. That's because ageing affects every cell of every organ, and your skin, your largest organ, reflects cellular damage in the form of wrinkles, dryness, dullness and discolouration. These are visible signs of ageing, reflecting poor health in need of repair. Invisible signs of ageing can affect your heart, your lungs and your brain with no outward manifestation.

Outer health and inner health are tied together in an infinite loop: the healthier you are on the inside, the healthier you are on the outside, and vice versa. So when you boost autophagy – which you do throughout Glow15 – it allows every cell to perform at its best, every organ to function at its best, so you look and feel your best.

You can help the autophagic process in many ways; that's what Glow15 is all about. But one of the most profound ways to activate autophagy is by nudging your body toward purity.

What does that mean?

Purity, to me, means that you treat your body with the respect it deserves, by warding off toxins internally and externally.

I think that perhaps the word "purity" can send the wrong message – that the goal has to be perfection, and if you skip a day of exercise or eat too much sugar or pull an all-nighter, you're ruined. That's not the case; you will not be ruined. You will not be less than. You will not be perfect, but that's OK, because you shouldn't be.

Instead, I want you to think of purity as the ability to prod, nudge and shift your behaviours – and your body – into the direction of acceptance, confidence, health and wellness. Purity is about living a life that feels good, happy, strong and young. After all, "glow" isn't just about dewy skin and looking beautiful – it's about *feeling* beautiful. Inside and out.

It's about a radiant life – no matter from what angle you see it.

Principle 2: Small Changes Yield Big Results

From an early age, I've been fascinated with numbers. I remember being seven years old and going for walks with my father, who as a chemist and mathematician always loves to share his view on numbers. We would go for an hour-long walk and count our steps; he would teach me basic algebra with apples we had just picked from the tree. We did multiplication tables over and over. When I was twenty-five, my father gave me the book *The Magic and Beauty of Numbers*. I love the way numerals look, I love the way they work, and I just love the majesty and science that come with formulas and functions.

I tell you this not to scare you into thinking I'm about to unveil some tricky statistical analysis for how to integrate autophagy into your life. Quite the opposite, actually.

I want you to embrace one mathematical equation:

$$1 + 1 = a \text{ lot more than you think!}$$

One of the frustrations many of us feel is that we don't *do* enough – that in order to have big effects, we have to make big changes (the superwoman complex). We have to turn our lives around 180 degrees. And because of that, we believe small changes won't make enough of a difference. We don't see immediate results, and then stop doing whatever it is we were doing. We don't give ourselves a chance.

That's the beauty of autophagy. Because it has such profound effects, you can make small changes – and your body will start experiencing big results. Some of those results may be noticeable quickly, while others may be happening at a cellular level and not as noticeable.

Glow15 will give you simple nutrition, exercise, sleep and beauty modifications. These changes will take a minimal amount of time.

I don't want you to think of your actions in absolute numbers (i.e., if you consume X fewer calories, you will lose X pounds). Instead, assume that every change you make has rocket fuel in it – with positive impacts felt throughout your body. This is what makes autophagy such an efficient and powerful biological process.

Now, I do want to briefly mention that dreaded five-letter word: *change*. Change, no doubt, can be difficult. That's the dilemma of any lifestyle switch: Can you adapt to a new way of living that gets you out of your normal habits and into newer ones? Can you get past the fear of change in your mind to make a change in your life? Certainly, how you feel about change depends on a number of factors, like the difficulty of the plan, the support you're getting from others, as well as your own boldness.

GLOW15 SUCCESS STORY

"I'm a better version of myself" MARGARET

"I think I spent the first week of Glow15 feeling guilty. I had a really hard time putting myself first – whether it was my food, my workout, my sleep schedule – it was tough not to feel like whatever I did was at the expense of someone else. I run our family business and feel responsible for my parents not only emotionally, but monetarily because I know it's their only source of income. My three kids, ages 16, 11 and 2, all need things at different times, and my husband, he helps, but I'm the one constantly doing everything for everyone else. Still, just like I am committed to my family, **I committed to Glow15. And the most amazing thing happened: I became a better daughter, a better mother and a better wife. I was more engaged. I was calmer. I was a better version of me.** This didn't happen overnight. The workouts were hard, but they slowly got easier. I stopped feeling like I just needed to make it through and started reveling in the challenge of each new exercise. The Low days were tough, and sometimes still are, but I like knowing the High day is less than 24 hours away. And I'm seeing the difference. My clothes are looser, and my body is tighter. I lost 6 pounds and 2 inches off my waist in just 15 days, And I literally see a difference when I look in the mirror. I have rosacea and my face is normally red and splotchy. Now, like my demeanor, it's calmer. I need less makeup. **Glow15 has helped me to glow – inside and out.**"

Many times, we get so frustrated with instructions and rules and "don't do" lists that we give up, because it's hard and we feel miserable. Our lives are hard enough already — and when we add the daunting task of making lifestyle changes, we think the pain and effort are not going to be worth it. After all, if it's no fun while we're doing it and it's taking a long time to see results, why bother?

Glow15 doesn't work that way. It incorporates small changes to ensure manageable adjustments. It's designed so that you won't feel the need to quit before reaping the benefits. One of the most telling parts of the clinical study (see page 23) was the fact that 97 per cent of the women stuck with the plan. They enjoyed it. They found value in it. They liked the new flavours and new habits, and they were better off for it – so much so that they still check in to report their continuing success. My hope is that you will do the same.

That, ultimately, will be what makes your fifteen-day jump-start become a long-term lifestyle.

Some women in the study did report that there were parts of the plan that were temporarily uncomfortable, because they were doing things they hadn't done before, like intermittent fasting or different kinds of exercise. But once they did those unfamiliar things for a bit, they learned to love them – which is the way it should work. Change can be intimidating, but I believe you will be so taken with how Glow15 makes you feel that the changes you make as part of the plan will be transformative. Knowing that small changes can lead to big results makes it much easier to push through the temporary discomfort – and keep your eyes on the ultimate prize.

You have the power to change your health. You have the power to change your life.

Principle 3: Less Is More

If only I had more time . . . for myself. If only I had more time . . . to cook healthy meals for my family. If only I had more time . . . I could exercise. If only I had more time . . . I could achieve the body, health, skin and life that I want. If only I had more time . . .

That's the way I once thought.

But I was wrong. The answer isn't to wish for more time.

The answer lies in working with the time you already have. Glow15 values simplicity and believes more can be accomplished with less. More reward for less disruption.

When you maximize autophagy, less is more. And less is the answer to making cellular switches to restore your youth. When there is less toxic waste in your cells from the Inevitable and Accelerated Agers, your body will reap more benefits, like boosted immunity, protection against disease, and increased energy.

On Glow15, less is more means you can do shorter workouts to boost longevity. It means your meals can be prepared quickly and easily – taking less time, but with maximum amounts of pure and powerful foods to fight the signs of ageing. And it means the simple supplements suggested will give you the greatest nutritional value to make your body and brain work more efficiently. It also means you don't even need as much sleep as you might think – fewer hours of sleep may actually give you better sleep quality.

I believe the most efficient path is the only path. You will not sacrifice any time doing this plan; in fact, depending on what you do now, you may even save some time with its simplicity. That's the beauty of Glow15.

To prove my point, I'm keeping this principle short, because I know you'll want to spend less time reading and more time putting Glow15 into action.

Principle 4: Timing Matters

Much of what you'll do on the Glow15 plan is about *timing*. Timing, of course, isn't time *management*. Timing is about aligning actions with a certain moment to maximize effect. What you do is important, but *when* you do it is just as important.

That is a core part of Glow15 – when you eat, when you exercise and when you sleep will help you optimize autophagy and get the results you want. That's because autophagy is influenced by your circadian rhythm, your natural waking and sleeping cycle. Think of it as a symphony: your brain is the conductor, your body is the orchestra and your circadian rhythm sets the beat, helping your body to stay in time. You can make physiological, mental and behavioural changes by understanding your body's 24-hour clock.

The autophagy-activating actions you take sync up with when you do them through that cycle. Recent research has shown that autophagy is rhythmically activated by our internal clocks. And that cycle is key to making it work efficiently.

To that end, you will find that cycling is a major theme of Glow15. Cycling may be something you're not used to. Rather than forming a habit and doing the same thing every single day, you'll be creating a rhythm of turning autophagy on and off.

The idea is that if you're constantly in a state of having your cells just do the jobs they're supposed to do, they don't have time to renew and rest and clean themselves out. You have to strike a balance – turning on autophagy because it naturally slows down as we age and because it's naturally turned off during our normal routines.

This on/off cycle provides a natural rhythm to your body's cells. Considering the way I've talked about autophagy, you may think you want it revving around the clock, but in your body, as in life, you can have too much of a good thing. Calories are designed to help us by providing energy to our bodies so we can function, but what happens if we have too much of a good thing? Those calories get stored as fat, which can have a host of negative consequences. Think about your immune system: it's designed to fight off invaders of all sorts – bacteria, viruses and more. But sometimes an overactive immune system can destroy healthy cells; that's what autoimmune diseases are – a form of friendly fire. That's the way to think about autophagy: it's a process that declines when we age, so we need to bolster it, but we don't want it activated 24/7. When we achieve that balance, activating autophagy in reasonable amounts, we'll find the ultimate life boosts.

Think of it as exhaling and inhaling. We can't do one without the other, and it's the back-and-forth of this process that makes breathing work. We'll be doing that here, too – activating and deactivating autophagy. But instead of doing it with your lungs, you'll be doing it with slight adjustments to your day.

In lots of ways, timing is a funny thing. Most people will chalk up "timing" to some supernatural force. Coincidence. Luck. Fate. Yes, that certainly seems to be the case many times. But when it comes to our bodies, timing is not luck. It's strategy.

Principle 5: Make Glow15 Your Own

I once read that to "break the rules you must master them". That's how I want you to think about Glow15. I'm going to get you started with specific instructions and then you're going to take them, learn from them and build your own way of living the Glow15 life.

Learning any skill is about a set of progressions – you start with the basics that act as a foundation you build upon until, eventually, you're doing your own thing.

Take a look at music in particular. My daughter plays guitar, and my son plays the piano. They learned those skills the way anyone does: by first learning the notes and scales and practicing them to get a sense of how everything works together. They practice the parts, and then they put them together. They worked their way up to songs, and if they keep on going, they're going to use the skills they've learned to do even more – maybe create their own songs, maybe improvise within songs, and definitely add their own style and voice to whatever they play. They will do what any musician or artist does: take the tools we all have access to and then make them their own.

That's exactly what I want you to do with Glow15: learn the notes and scales – and then feel free to improvise and make it your own. It is my hope that you take the main tactics and principles of Glow15 and create your own symphony of sorts.

I picked fifteen days because that's enough time for you to see some powerful and motivating results, be inspired and learn to start forming new habits. It's also enough time to see what notes you like to play and how you can riff – to make the plan your own. There's also evidence, according to a study published in the *Journal of Molecular Medicine*, that autophagy can be accelerated within fifteen days, which can give you that kick-start to see some of your work in action.

Like any biological function, autophagy will work differently in every

GLOW15 SUCCESS STORY

"I did it for my daughter" SARAH

"When I first began Glow15, I was keeping a secret. I hid it from friends and coworkers and, most significantly, I kept it from my teenage daughter. I could never let her know how much I hated the way I looked and felt – fat, ugly and disgusting. And worse, my daughter, BaileyAnn, could never see how it affected my self-esteem. In my mind, this was not supposed to be important, this was vain. I'm the funny, smart one – and that's more valuable than being the pretty one, or that's what I wanted her to believe, anyway. But how could I tell BaileyAnn that I want her to have my heart and not my looks? Then **I started Glow15 and things changed.** First, I had a plan – something that told me what to eat, how to sleep, when to move and what to use. And secondly, I realized that the best way to influence how my daughter feels about herself was to start taking care

of myself. For the first time since BaileyAnn was born (that was 16 years ago!) – I put me first. I never really thought about the emergency info you get on an airplane – 'help yourself before your child' – but all of a sudden, it made sense. Slowly, I started helping myself. I began to follow the plan – and I even started a skincare routine. This may not sound like a big deal, but up until now, I've used soap and water – that's it – to care for my skin. And to my surprise (and delight!) people noticed: my husband started giving me more compliments – he told me I glowed the way I had the day I walked down the aisle. **My coworkers wanted to know what I did or where I went that left me so 'radiant'.** But the best was, and is, BaileyAnn. Not only did she notice a change in how I looked, but, for the first time in her life, she knows there are no secrets about how I feel."

person. I designed Glow15 for you to customize it into your own unique approach to attack accelerated ageing. And when you're done with those fifteen days, my guess is that you'll be silently saying to yourself, *Encore! Encore! Encore!*

Now let's see how the big-picture principles turn into the strategies that you will use on Glow15 when it comes to the major areas of your health: nutrition, exercise, sleep, and beauty.

Chapter 4

Eat, Drink, Glow

Nutrition to Make Your Cells Younger

The words detox and cleanse are used so often that, to many, they have become meaningless. You've seen them in every magazine. You've heard them on the news. You know they are used to tout juices and spas and all things healthy. But while these words are often used to promote healthy behaviour, nothing they are being used to promote will actually flush out toxins from your cells.

That's OK, because you naturally have the ability to get your body to "detox" and "cleanse" itself. And *you* can control that process.

All you need to do is get your body to eat itself – eat away your wrinkles, eat away your exhaustion and eat away brain fog – and, yes, eat away your body's own fat, too! Remember, *autophagy* literally means "self-eating" and it is a natural detox process that allows for the removal of toxins from your cells.

So how do you get your cells to self-cannibalize? The answer, strangely enough – and yes, I get that this is strange – is in what and when you feed them. Understanding how nutrition can activate autophagy gives you the power to detox your cells and slow ageing.

Glow15 is not a diet. A diet is all about numbers: the number on the scales, the number of calories you eat and burn, and the number of foods you're allowed to eat. On a diet, success is defined in terms of how well you stick to your numbers. Instead, this plan is all about results: energy, vitality, weight loss, beauty and strength. I'll give you the guidelines and nutrients to add to your existing food plan for maximum youth benefits. On Glow15, success is defined in terms of how well you feel, look and thrive.

In this chapter, I will take you through the Glow15 way of eating. Backed by award-winning scientific research and made easy to follow by leading nutritionists, it is designed to be cyclical, turning autophagy on and off. You'll find that in a very short time, it will become as easy as breathing in and out.

Guidelines to Get You Glowing

So how do you get your body to eat itself? Autophagy is activated in "stress mode," so you'll want to stress your cells a little bit in order to get them to start dining on your fat, wrinkles, brain fog and exhaustion. One of the best ways to do this is through smart nutrition.

Smart nutrition starts with understanding that there's value in all the macronutrients (protein, fat and carbohydrates) and micronutrients (vitamins, polyphenols and minerals) found in natural foods. Again, instead of getting into a laundry list of "do" and "do not" and "don't you dare" foods, it's best to think about what foods will do the most to boost your youthfulness. It's also important to learn more about timing as it relates to your diet, because when you eat can make the biggest impact on how youthful you look and feel.

1: IFPC

The acronym may sound like it stands for some kind of international health organization, but IFPC serves as the Glow15 approach to eating to activate your youth. IF stands for intermittent fasting, and PC stands for protein cycling. While it may sound complicated, it's not in practice.

Together, IF and PC serve as your autophagy on/off switch, the ignition that gets autophagy going. As with any fire, you need the spark to start it – and IFPC is it. As your body cycles through IFPC, you create a rhythm to your nutrition that allows autophagy to do its job – and this makes your cells act younger and healthier than they actually are.

Let's take a look at how IF and PC work separately and then what happens when you put them together.

Intermittent Fasting (IF): IF is the practice of shifting between periods of unrestricted eating and restricted eating and is a key activator of autophagy. This is a natural and extremely effective way to put your cellular cleansing cleanup crew to work. If you're constantly eating – which is the case for many of us as we graze throughout the day – it doesn't give your cells a chance to repair and clean up the waste and toxins they have accumulated. Short periods of not eating give them the time to take care of those tasks.

Specifically, IF works by activating glucagon, which works in opposition to insulin to keep your blood glucose levels balanced. Think of a seesaw: if one side goes up, the other goes down. In your body, if insulin goes up, glucagon goes down, and vice versa. When you give your body food, insulin automatically rises and glucagon starts to decrease. But the opposite happens when you deny your body nutrients – insulin goes down and glucagon rises. An increase in glucagon triggers autophagy. This is why temporarily withholding nutrients, or intermittent fasting, is one of the best ways to boost the youth of your cells.

Research has shown that many of the benefits of IF – like burning more fat, providing more energy and decreasing your risk of developing diabetes and heart disease – can be attributed to the activation of autophagy.

A fascinating finding from the journal *Obesity* is that in many cases, IF encourages the body to burn more fat while sparing muscle. In fact, it is four times better than caloric restriction at helping your body burn fat while preventing lean tissue loss.

In practical terms, here's how IF works on the Glow15 plan: on three nonconsecutive days (your Low days), you will intermittent fast for 16 hours and eat during an 8-hour period. You can begin your intermittent fast after dinner the night before so the majority of your fasting hours occur during sleep. For example, if you stop eating at 8 p.m. and fast overnight, your first meal will be at noon the next day, so essentially, you'll be skipping only breakfast. As long as your intermittent fast lasts for 16 hours, you can adapt your IF to your schedule. If you prefer to eat breakfast, you can start your intermittent fast earlier by skipping dinner the night before. Make the hours work for you, your body and your lifestyle.

There is no need to intermittent fast for any longer than 16 hours. Research shows that 16 hours is optimal for creating the caloric restriction that happens during intermittent fasting and can activate autophagy through different nutrient pathways. A study published in the journal *Cell Metabolism,* led by Valter Longo from the University of Southern California with the participation of other researchers and clinicians from around the globe, reported that cycles of intermittent fasting have numerous positive effects, such as decreasing visceral fat, reducing cancer rates, improving the immune system, slowing down the loss of bone mineral density and increasing longevity.

Protein Cycling (PC): PC is the practice of alternating between periods of low protein consumption and normal to high protein consumption. On Glow15 you will limit your protein intake to about 25 grams on the same three days you practice IF (your Low days). The other four days (your High days), you will eat normal to high amounts of protein.

PC has an effect similar to fasting. Creating protein deficiency also lowers your insulin levels – it's that seesaw again – and that, in turn, boosts your glucagon and activates autophagy. This means your body will not store the foods you eat as fat, but instead work to build muscle and burn fat.

One of the main reasons PC works to enhance youth is because your body can't create its own protein. Instead, it is forced to find every possible way to recycle the existing protein you've already provided it. If you deprive your body of protein, it will enhance autophagy, kicking your body's recycling programme into overdrive.

The problem is, we don't normally deprive ourselves of protein. Actually, we generally eat enough protein to keep autophagy in "maintenance" mode. On average we eat 70 grams of protein daily, which is more than one and half times the amount typically recommended for women. But our bodies can handle periods without protein – if you think about it, this goes back to our ancestors, the hunter-gatherers who often had to survive for periods without a successful hunt. (Interesting note: Breast milk, the ideal and most complete food for babies, meant to be consumed when tiny bodies and brains need the best possible nutrition and are growing the fastest, derives only 6 per cent of its calories from protein.) Now we have to *choose* to deny ourselves protein.

To be clear, low is not always the way to go. Being in a constant state of low protein will actually contribute to ageing in the form of muscle wasting, accompanied by increasing weakness and immune deficiencies. So you need to have both high and low protein days – that's the cycling part. There is evidence that protein cycling can help reduce the risk of diseases, diabetes, cancer and heart disease, in addition to enhancing autophagy.

IFPC: Picture the ocean. Those waves, big or small, are mesmerizing as they roll up onto shore, then retreat back. Up and back, up and back. High tide and low tide. Strong, then soft. The beauty of the ocean comes from that serene rhythm. Autophagy – like the ocean – is driven by cycles and rhythms. It needs to have an ebb and flow, which is what IFPC allows your body to do – rolling in and out of autophagy cycles. Remember, you want to turn autophagy on and off – inducing autophagy without ever inhibiting it is like always inhaling without ever exhaling.

When autophagy is turned on by intermittent fasting plus protein restriction, it creates a true shortage in your body that can be corrected only by your cellular cleansing cleanup crew doing its job. Of course, eventually the restriction must end to prevent overworking the system and damaging your cells, but for the entire restriction period, autophagy will be at work, since human cells must constantly make new proteins, regardless of conditions, in order to function.

Research shows that withholding specific nutrients from your body is actually better for your metabolism than continuous caloric restriction: markers of metabolic health experienced greater improvements, including improvements in cholesterol, blood pressure and inflammation, from intermittent fasting and protein restriction than from calorie restriction alone.

On Glow15, your three nonconsecutive days of fasting and protein restriction will be your Low days, and the four remaining days will be your High days. These can be any days you choose, but I like to be specific, and the majority of the women on the programme picked Monday, Wednesday and Friday as their Low days so they could indulge in unrestrained eating on the weekend.

2: Fat First, Carbs Last

Remember that one of the key principles of Glow15 is "timing matters" – *when you eat what* is more important than *what you eat when*. This is especially true of when you should have nutrients like fat and carbs to best activate your autophagy.

To allow your cellular cleansing cleanup crew to work efficiently, start your day, your first meal, with fats first and save your carbs for later. Let's take a closer look at why the timing of these foods is critical in defying ageing.

For years, dietary fat was portrayed as the primary adversary to our health. *Fat is bad for you. Fat is evil. Fat will, well, make you fat.* That was the hallmark line of thinking during the "low-fat diet" craze – a movement that saw food manufacturers remove fat from products, only to replace it with loads of sugar and other junk that was intended to fill the flavour void left by fat (they had to find a way to make their low-fat and fat-free products taste good, after all). In the end, that's what messed up our bodies so badly: too much sugar, not enough fat. We all saw this push in supermarkets and restaurants and everywhere we looked, with all those "fat-free" or "reduced-fat" labels and products (which stimulated cravings and led to overeating). Well-intentioned people passed along this message and this practice by saying that it was a means to better health. But they made a major mistake.

While many of us have been conditioned to think of dietary fat as our enemy, this is simply not the case. The truth is that fat does not clog our

High Days (4 days a week)

Unrestricted eating (no time window)

Eat average to high amounts of protein (10 to 35 per cent of daily calories, or 45 to 157 grams)

How to calculate higher protein Multiply your weight (in pounds) by 0.36 (grams of protein recommended per pound of body weight).
For example, for a 130-pound woman: 130 × .36 = 46.8, or approximately 46 grams of protein

Low Days (3 nonconsecutive days a week)

Fast for 16 hours (followed by 8 hours of eating)

Eat low amounts of protein (5 per cent of daily calories, or less than 25 grams)

How to calculate lower protein Multiply your weight (in pounds) by 0.36 (grams of protein recommended per pound of body weight) and divide by 2.
For example, for a 130-pound woman: 130 × .36 = 46.8 and 46.8/2 = 23.4, or approximately 23 grams of protein

arteries, it does not raise cholesterol and it does not make us gain weight. Dietary fat isn't a villain. In fact, fat is our friend – it's absolutely essential to optimal health. Fat is your body's preferred fuel. It's satiating, making it protective against cravings for unhealthy foods. It's a nutritional hero with incredible powers to help you achieve optimum health.

My mantra is "fat first, fat most, everyday in every way".

And, notably, autophagy is promoted by fat. Studies especially link fats like MCT (medium-chain triglyceride) oil and omega-3 fatty acids to the promotion of autophagy. These natural, unprocessed fats are found in real foods like avocado, butter, nuts, salmon, omega-3 fish oil, coconut oil and avocado oil. Whereas eating protein or carbs can trigger insulin to be released – which, if you recall from the explanation of IFPC on page 56, works counter to autophagy-activating glucagon – fat does not. It does not promote the release of insulin. Instead, fat can prevent erratic spikes in blood sugar as well as keep you satiated.

Thankfully, I grew up with fat as the most abundant food in every meal or snack I ate. My parents never bought low-fat anything – ice cream, yogurt, crackers or cottage cheese. They believed in full-fat foods and saw low-fat foods for what they were: overprocessed rubbish that was masquerading as healthy food. My four children start every day the same way I did – by eating full-fat breakfasts. My eldest child, Megan, has started with fat first since she was in a highchair. Avocado was the first solid I gave her, and she continues to eat it to this day. Luckily for me, the rest of the children like to follow in their big sister's footsteps, and I knew I was setting them up for a healthy life.

When You Eat Fat: This is nearly as important as the intake of fat itself. Eat fat first and make sure you incorporate it as part of every meal. A study in the *International Journal of Obesity* found that mice fed a high-fat meal at the beginning of the day had normal metabolic profiles, in contrast to those that ate a carbohydrate-rich diet in the morning. The latter resulted in increased weight gain, body fat per centage and glucose intolerance, along with other metabolic markers. Fat intake at the time of waking can kick-start fat metabolism and keep you satiated throughout the day. This is why you will always have fat first on Glow15 – regardless of whether you are on a High day or a Low day, the first thing you eat or drink will be full of fat. I start every day with AutophaTea (more on that on page 65). Adding fat to my morning tea provides satiety without adding more protein or carbs that could cut into my daily allowance on Low days – and it tastes good every day and curbs my cravings! Along with my tea, I like to eat an AvocaGlow (page 200) or one of my Egg15 muffins (pages 201–216).

On my Low days, I eat around noon, and on my High days, I'll indulge soon after waking. Also, on my High days, I often like to add a nitrate-free bacon, which has some protein, but also gives me some additional flavourful fat. Perhaps one of the most interesting and helpful reasons to have fat first is that it carries the flavour in the food. The more fat you eat, the less sugar and salt your taste receptors require to register satisfaction. We're biologically wired to feel pleasure from food, and fat is the macronutrient that most satisfies this need for pleasure.

Activating your autophagy with fat first not only gives you sustained energy, regulates your metabolism and fights the signs of ageing, it can also hydrate your body. One by-product of metabolizing food is water. More water is released when we consume fat than when we eat carbohydrates or protein – which means fat can help you glow.

What Kind of Fat You Eat: Embrace some of the good fats that the world has. They are found in fish like mackerel, salmon and sardines, in olive oil and avocado oil, in nuts like macademia, in seeds like flax and sunflower, and in avocados – all important, youth-boosting foods that are part of Glow15. (For more details on the types of fats to use and when and how to use them, refer to chapter 9.)

The only type of dietary fat you should not include in your diet is trans fat. Trans fats are artificial and are created when hydrogen is added to liquid vegetable oil to make it solid at room temperature. These must be avoided.

Carbs Last: Again, when you eat is important on Glow15, and this is particularly important with carbohydrates. It's why you will not have carbs in the morning, but instead save them for later in the day.

The reason for this is that ketosis – a natural metabolic state in which your body uses fat as fuel in the absence of carbohydrates – is one of the best ways to boost autophagy. "Ketosis is like an autophagy hack," says Dr Colin Champ, an assistant professor at the University of Pittsburgh Medical Center. "You get a lot of the same metabolic changes and benefits of fasting without actually fasting." However, on Glow15, your body does not need to be in a constant state of ketosis to reap its benefits and activate your cellular cleansing cleanup crew. This is because when you first wake, your body is already high in ketones. By skipping carbs for at least the first half of the day, you can take advantage of the autophagy that has already been initiated and prolong this nearly ketogenic state to further benefit from it.

Saving your carbs for later in the day also allows your body to use them most effectively. Carbohydrates can make your seesaw go haywire because they promote the release of insulin. When your blood sugar soars and plummets, you'll feel tired and sluggish. But you need carbs to function – plus, they're delicious! Carbs help bring on that stress-response mode, encouraging autophagy, and they make up some wonderful, autophagy-maximizing foods. Carbs also help provide energy for your body and help relax you, so eating your carbs later in the day can help you prepare for sleep.

You can enjoy all types of carbs on Glow15; however, carb *quality* matters.

To ensure you're getting the highest-quality carbs, there are a few tips and tricks that I've found very helpful. First, eat "whole" carbs instead of processed or refined carbs. Examples include vegetables, fruits, legumes and grains. Whole carbs contain higher amounts of fibre, making them a high-quality carb. Using Harvard University's 10-to-1 rule can help you determine if a packaged food is high in fibre. Look at the label to find the total grams of fibre and multiply by 10. Then, compare that number to the total grams of carbs listed. If the amount of fibre multiplied by 10 is equal to or higher than the total grams of carbs, you have a high-quality carb.

How you prepare your carbs can also play a role in autophagy. For example, al dente pasta is better for autophagy than overcooked pasta. A green banana is better than a yellow one. And white potatoes that are cooked and cooled, like in potato salad, have much more impact on autophagy than mashed potatoes. Why? Scientists are discovering that an important fibre called resistant starch is a key player in our digestion as well as in the hormones regulating blood sugar.

By always starting your day with fat first and ending with carbs last, you can reap the cell-renewing, life-enhancing and -extending benefits of autophagy every day, not just on your Low days. You get benefits similar to IFPC by regulating the timing of two macronutrients – fat and carbs – and eating until you're full. I believe this makes Glow15 enjoyable and feasible enough to practice for an especially long and healthy lifetime.

3: The Ultimate Youth Boost: AutophaTea

There's a good chance that you already have a morning pick-me-up. For many of you, it's a cup of coffee or tea. For others, it might be diet cola. Or maybe you're someone who reaches for an energy drink. Maybe you choose these beverages because you're sluggish from a poor night's sleep (see chapter 7 to improve sleep), or maybe you do it because it's habit. But what if I told you there's something better than all those options?

What if your morning pick-me-up could do more than just give you energy? What if it could also help you burn fat, boost your brain power, suppress hunger, increase your immunity and make your skin glow?

That's exactly what AutophaTea does.

I wanted to find one simple way to maximize autophagy every day, using the most potent autophagy-activating ingredients and making it easy to start your day with fat first. Plus, it had to taste delicious, be simple to prepare, and have the same top-of-the-morning juju that other caffeinated drinks provide.

I discovered the key to this age-defying drink when I was in Calabria, Italy, to meet with Dr Elzbieta Janda and other scientists to learn about the powerful cholesterol-reducing bergamot fruit. We started our exploration at the birthplace of bergamot, on the sun-drenched slopes of the Aspromonte mountain. The combination of the sea, soil and air all come together to form unique growing conditions for this amazing fruit. Together we began to climb the mountain to meet with the farmers who grow these rare and potent fruits in their orchards. There we met Hugo, a fourth-generation farmer, who picked the fruit by hand as his father, grandfather and great-grandfather had done before him. As we walked through the orchard with him, Hugo carefully plucked a fruit from one of his trees and gently cut it open with his penknife. He made sure to include the pith – the thin white layer below the rind. This was critically impor-tant, Dr Janda said, as the pith holds the majority of the fruit's powerful polyphenols. We each took a bite: the bergamot had a delicious, fresh, sweet-and-sour taste.

Back at Dr Janda's laboratories, she explained that the juice and the pith were strong autophagy activators that could slow ageing. While we were speaking, she offered me a cup of tea made from the fruit. Of course, I said yes – I grew up in England and have always loved drinking tea. Dr Janda's tea was a unique version of my favourite Earl Grey, created using the whole citrus bergamot. That afternoon we drank three cups of tea each, and I had to ask Dr Janda why so many. She laughed and said that since her life's work was dedicated to researching autophagy, she had to make sure she was activating her own so she could live younger with lots of energy to be the poster child of her work.

I was so excited about what I learned from Dr Janda that when I returned home, I brought together my research team of PhDs, doctors and nutritionists, and even had some tea experts weigh in. We realized we could add the whole citrus bergamot fruit to Earl Grey to make it a more effective youth activator. We identified three additional important and powerful ingredients to amplify autophagy's anti-ageing effects: green tea, a powerful polyphenol that wakes up your metabolism; cinnamon, an immunity booster; and coconut oil, which is full of good fat to keep you satiated and give you energy. We mixed the three ingredients with the whole citrus bergamot fruit Earl Grey and, from the very first cup, I was hooked!

I could feel its power – I was energized, my mind was clearer and I was satiated in a way I had never experienced from a drink before. Plus, it was delicious. I wanted to see if other people felt the same way I did, so I had my friends and family (my own personal guinea pigs) brew some of their own. Overwhelmingly, they agreed with my findings. Then we tried it out on the women in the Jacksonville University Glow15 study. Overwhelmingly, they agreed, too. In fact, more than 80 per cent of the women who followed Glow15 reported that the AutophaTea was their favourite, tastiest and easiest part of the plan. I heard over and over again that it was delicious, and I liked that so many women described it as drinking a warm cinnamon roll – an enjoyable and indulgent drink.

Plus, it came with all kinds of benefits. In fact, there's powerful science that shows that the ingredients in AutophaTea work to boost energy, burn fat faster, reduce anxiety, increase metabolism, sustain heart health, increase immunity, protect your brain and fight the diseases of ageing. Why is it so effective? It all comes down to the main ingredients.

Whole Citrus Bergamot Earl Grey Tea: Using the whole citrus bergamot fruit and oil, this polyphenol-rich tea can induce autophagy to remove toxins from your cells and keep your heart healthy. By doing so, it improves insulin sensitivity to lower blood sugar. With its statin-like properties, it can decrease total cholesterol and triglyceride levels. Earl Grey is a black tea and contains 40 to 60 milligrams of caffeine per cup. Many Earl Grey teas contain synthetic bergamot oil, which will not give you any of the autophagy benefits, so check the ingredients and look for Earl Grey tea made with bergamot oil, not bergamot flavouring.

Green Tea: This powerful polyphenol increases metabolism and reduces anxiety. Active polyphenols like EGCG (epigallocatechin gallate) in green tea activate autophagy. EGCG increases thermogenesis (the rate at which calories are burned), and studies show that people who drink green tea burn up to 100 calories more per day than those who don't drink it. Green tea leaves are heated immediately after plucking, which prevents them from oxidizing. This lack of oxidation is responsible for the very low caffeine content of green tea – only 1 per cent.

Cinnamon: This spice, high in antioxidants, has been shown to increase autophagy and have the ability to turn on longevity-related genes. Like green tea, cinnamon activates neuroprotective proteins and can reduce the negative effects of oxidative damage. Cinnamon has been shown to protect brain cells from neurodegenerative diseases like Alzheimer's, Parkinson's and dementia. Cinnamon can also protect against cell damage, cell mutation and cancerous tumor growth. Studies have revealed that these health benefits come from an autophagy-inducing compound called cinnamaldehyde, which can inhibit cancer tumor growth by protecting cells from damage, while also encouraging cancerous cells to self-destruct – a process known as apoptosis that works in conjunction with autophagy. Not all cinnamon is created equal: look for Ceylon cinnamon, rather than the more commonly found cassia or Chinese cinnamon, as it is considered "true cinnamon" and is of the highest quality.

Coconut Oil: Coconut oil is nature's richest source of healthy medium-chain fatty acids, making it the perfect ingredient to help you start your day with fat first. It eases your hunger and stimulates autophagy by increasing ketone levels – especially in the absence of carbs. This combo is ideal for getting your cellular cleansing cleanup crew to work, as the MCT in coconut oil permeates cell membranes easily and can be utilized effectively by your body. It's easily digested because the MCTs are sent directly to your liver, where they are immediately converted into energy rather than being stored as fat. Also, the ketones produced stimulate metabolism, and that promotes fat burning.

Every day on Glow15, you'll start with a cup of AutophaTea. On High days, you'll have it first thing in the morning, and on Low Days, you can drink it to break your fast. Each day, you can drink up to four cups, though I suggest you switch to decaf versions or stop drinking it after 2 p.m. because of its caffeine content. (You don't have to give up coffee, if you like that, since coffee activates autophagy. Just have your AutophaTea first.)

Cheers to your youth!

4: Discover Autophagy-Activating Foods

Ever heard of saponins, sphingolipids or sulforaphanes? Probably not. But they are all compounds found in foods shown to aid in autophagy. While these names may not be familiar, many of the foods are probably already in your kitchen.

The Glow15 way is not about elimination, but about embracing the wonderful foods you can eat. I will provide easy-to-prepare and delicious recipes that take the guesswork out of the equation, helping you activate your youth. Here's a closer look at the best autophagy-optimizing foods.

Polyphenols: These antioxidants protect against free radical stress and damage to cells. Eating a polyphenol-rich diet will activate autophagy to produce similar effects as IFPC in regulating life span and your metabolism. And some of the best sources of polyphenols are also the most indulgent. For example, the polyphenol resveratrol, found in red wine, has been shown to activate autophagy to reduce inflammation, boost heart

GLOW15 SUCCESS STORY

"Healthy AutophaTea replaced my unhealthy cola"

TIFFANY

"I never thought I'd be saying this – but it *is* possible to live without cola! I've been a cola drinker since high school – I drank it to get going in the morning, I drank it to pick me up in the afternoon, and I drank it with my meals and for snacks between. My doctors told me to quit – they were worried about my blood sugar and cholesterol. I knew it wasn't good for me, but I liked the taste and the energy surge. I started Glow15 with the intention to get healthier, so I decided that I would replace my soda with AutophaTea. I was totally sceptical – tea with oil? – but it was surprisingly tasty. Remember cinnamon crunch cereal? I used to love drinking the milk from my bowl after I finished the cereal. AutophaTea tastes like a gourmet version of that cinnamon-y goodness. I drink 2 or 3 cups every day and I'm still impressed with how it keeps my hunger in check. It took a few days to really kick in, but then I found myself getting fuller faster – even on High days. **I lost over 7 pounds in the first 15 days and my body mass dropped by 2 per cent!** But for me, the best part of Glow15 is what I gained: energy and confidence. **I had more pep than I've had in years – a new 'get up and go' feeling! And my mind seemed clearer and more focused.** With two young kids who always seem to be on different sleep schedules, I never expected to feel this good. **I feel proud, I feel accomplished, and I feel I owe it all to Glow15."**

health and inhibit breast cancer. Go ahead, pour yourself a glass. And grab some dark chocolate, too — it's filled with polyphenols that can activate autophagy. It also protects against free radical destruction, increases fat breakdown, and helps stave off hunger.

Food sources for polyphenols include:

- Fruits: bergamot, black currants, blueberries, plums, prunes, strawberries
- Herbs and spices: basil, cloves, cumin, peppermint
- Nuts and seeds: almonds, flaxseed, hazelnuts, walnuts
- Vegetables: artichokes, asparagus, black olives, broccoli, capers, chicory, endive, leafy greens, red lettuce, spinach, tomatoes
- Other: red wine, cocoa powder, dark chocolate, coffee, green tea

Eat fruit raw and lightly cook vegetables to retain the most polyphenols. A study in the *Journal of Agriculture and Food Chemistry* also showed statistically higher levels of polyphenols in organic and sustainably grown foods compared to conventional varieties.

I believe that polyphenols are one of the best ways to supplement the Glow15 diet. In the next chapter, I'll detail my favourite types of polyphenols – called Powerphenols – to boost your cellular cleansing cleanup crew and fight ageing.

Sphingolipids: Think of a brick wall. Mortar is the sticky substance that holds the bricks together. As the wall ages and is exposed to the environment, the mortar can break down, causing the bricks to crumble and the entire structure to lose its strength. In your skin, sphingolipids are lipids in the cell membrane that act like mortar to keep the foundation of your skin strong. Sphingolipids decrease with age, and over time the cells of your skin are less able to hold their form – and that's what makes your skin thinner and weaker. This is why it's important to keep your sphingolipid levels high through the foods that you eat. The benefit? Fuller, plumper, hydrated skin. Beyond the skin, they regulate neurotransmitters binding to their receptors, important in healthy brain function (read: better moods, better decision-making, increased critical thinking and improved memory). They can also trick your cells into thinking you're fasting even on days when you're not, which helps promote autophagy.

Food sources for sphingolipids include:

- Carbohydrates: rice, rice flour products, sweet potatoes, wholewheat bread, wholewheat flour, wheat germ

- Fats: butter, cream
- Protein: cheese, chicken, cottage cheese, cow's milk, beef, eggs, pork, turkey, yogurt

Omega-3 Fatty Acids: These good fats help protect the liver from the toxic effects of bad fats by turning up autophagy, suppressing inflammation and stimulating fasting in some cells. Omega-3s have been shown to help improve heart health and serve an important role in protecting your brain cells. They work in part by preventing the misfolding of a protein resulting from a gene mutation in neurodegenerative diseases like Parkinson's.

Food sources for omega-3s include:

- Protein: bass, crab, pollock, herring, salmon, mackerel, sardines, cold water prawns, trout, fresh tuna, grass-fed meat, egg yolks from omega-3-rich eggs
- Nuts and seeds: chia seeds, flaxseed, walnuts
- Vegetables: Brussels sprouts, kale, spinach, watercress
- Oils: fish oil, flax seed oil, hemp oil, walnut oil

Sulforaphanes: These sulfur-containing compounds are released by cruciferous vegetables such as broccoli and cauliflower when you chew or cook them. Sulforaphanes initiate autophagy by turning up processes that degrade old, abnormal proteins. They also behave as an antioxidant, blocking free radicals from causing damage to cells. They help brain health by activating autophagy in neuronal cells. In addition, sulforaphanes boost autophagy in some cells in the eye, which can help with vision issues as you age. You should eat foods containing sulforaphanes raw, because some of their nutritional content can leach into cooking water. I like to top my salad with fresh broccoli sprouts. If you can't tolerate cruciferous vegetables raw, try steaming them for just two or three minutes.

Food sources for sulforaphanes include broccoli, broccoli sprouts, Brussels sprouts, cabbage (red or savoy), cauliflower, spring greens, horseradish, kale, and kohlrabi.

Vitamins C, D, E, and K: These vitamins improve the quality of cell-to-cell communication and help regulate autophagy, playing an important role in the prevention of cancer. Vitamin C is a powerful antioxidant that is currently being researched as a less invasive and more affordable cancer treatment. Vitamin D is crucial in immune system regulation. Vitamin E plays a role in cancer-cell death and may be useful for treatment in gastric cancers. And vitamin K_2 helps kill off abnormal cells and is touted for its anti-tumor effect.

Food sources for vitamins C, D, E, and K include:

- Fruit: whole citrus bergamot, strawberries
- Protein: almonds, cheese, hazelnuts, herring, milk, peanut butter, salmon, sunflower seeds
- Vegetables: spring greens, dark leafy greens, kale, spinach, turnip greens
- Other: cod liver oil, wheat germ

Spermidine: This natural polyamine is found both in plants and in human cells. In humans, levels of spermidine decrease with age, which is important because they directly interact with DNA and turn on genes that regulate autophagy. A recent study showed that spermidine can help increase life span and reverse age-related cardiac problems, and it's also been shown to help with memory. And as the name implies, spermidine is found in semen – but before this takes an X-rated turn, you can also find it in your local grocery store. Look for aged cheddar cheese, as it has one of the highest levels of spermidine.

Food sources of spermidine include:

- Protein: cheese, chicken, steak
- Vegetables: broccoli, cauliflower, lentils, mushroom, peas, potatoes, red beans
- Fruit: grapefruit, pears
- Other: soy sauce, vinegar, alcohol, fermented foods

Saponins: These phytochemicals promote an anti-inflammatory environment systemically. They activate autophagy by stimulating your immune system, making membranes more permeable to protein activators. Animal studies reveal the ability of saponins to promote balanced blood sugar. This is crucial to autophagy, as the presence of glucose triggers insulin secretion, which turns on genes that suppress autophagy. These soap-like compounds are usually bitter-tasting and foam when shaken in water.

Food sources of saponins include:

- Grains: amaranth, barley, brown rice, buckwheat, millet, oats, quinoa
- Legumes: black beans, chickpeas, broad beans, kidney beans, butter beans, peanuts
- Vegetables: alfalfa sprouts, aubergine, green beans, peas, peppers, potatoes, tomatoes
- Other: garlic, red wine

Probiotics: Probiotics are beneficial strains of bacteria found in fermented foods and as supplements. The use of probiotics increases longevity by suppressing chronic low-grade inflammation in the body. They help with digestion by creating substances that promote autophagy; they do this by cleaning out some of the bad digestive gunk in our body. When bad bacteria take over, your body can't repair your gut as well, opening you up to inflammatory problems like irritable bowel syndrome and Crohn's disease. Probiotics help your gut by sending cellular signals that turn on autophagy and remove that gunk from your cells.

Sources of probiotics include:

- Protein: kefir/acidophilus milk, tempeh, yogurt
- Vegetables: kimchi, sauerkraut
- Other: dark chocolate

You can find Glow15 recipes using all these autophagy-activating food groups starting on page 198.

5: Pure over Processed

You can probably guess where I stand when it comes to clean eating, since you know I grew up eating naturally and organically. I should say that as a teenager, I did what most kids that age tend to do: I strayed. Every single day on my way back from middle school, I bought a candy bar (sorry, Dad). But the foundation my parents built for me really programmed my body to eat the clean way; I just didn't feel good afterward because my stomach couldn't handle it (it literally made me sick). I felt good eating the foods that powered me and that made me feel energized and strong.

To this day, I believe that the fewer ingredients foods contain and the closer they are to their natural state, the better.

On Glow15, do your best to follow these pure food guidelines:

- Eat whole foods that are more a product of nature than a product of industry (like processed and packaged foods). Look for foods stored in glass rather than plastic.
- Use natural sweeteners like monk fruit and honey instead of artificial or refined sugars; drink AutophaTea or coffee instead of fizzy drinks (regular or diet).
- Beware of sugar code names, like dextrose, maltose, corn syrup, fructose, malt and maltodextrin.
- When possible, eat organic fruits and vegetables and grass-fed/ pasture-raised/free-range meat, free of pesticides and herbicides.

By following the Glow15 guidelines, you will see results – from your shrinking waistline to your growing energy levels – in just fifteen days. But making them a permanent part of your lifestyle – and further maximizing the effects with the Powerphenols outlined in the next chapter– will allow you to reap all the benefits of Glow15 for a youthful and long life.

GLOW15 SUCCESS STORY

"Pure over processed pays off" GRACE

"I first tried Glow15 because I bought a 'goal' outfit for my son's college graduation and I really wanted to look good. The yellow sleeveless top and blue pants were a size 12 but I'm a size 14. Or, I was a size 14. **In just 15 days, I reached my goal.** The pants buttoned easily and I lifted my bare arms and actually pumped a 'hooray' when the top slid down just skimming my new flatter stomach. I got so many compliments. Everyone wanted to know the secret to what I'd been doing. I told them all about Glow15 and as I did I realized that it helped me do a lot more than fit into my clothes. I'm in my 50s and had always started my day with heaped spoonfuls of sugar substitutes in my tea. But the Glow15 guideline 'pure over processed' gave me a new perspective on what I was putting in my body, so I drank the AutophaTea without any of my usual add-ins. It was really hard. I was tired, got headaches and I missed my old sweeteners. But after that, I was good – really, really good. I could feel a difference – **my cravings disappeared, my thoughts became clearer and my mood improved – and I lost 8.5 pounds!** I have stayed off sugar ever since and it feels great. **I thank Glow15 for helping me make a permanent change to better my health and my life – and making me look good too!** On the day of my son's graduation, when the sugar substitutes were handed to me, I passed. And instead, I shared the coconut oil and cinnamon sticks I now travel with, and got everyone hooked on AutophaTea."

Chapter 5

Powerphenols

Youth-Boosting Nutrients to Protect and Repair Your Cells

When I was a little girl, my mother would lay out my breakfast of eggs and berries with a side of cod liver oil. I know, you might think, *Ewww, cod liver oil,* but the truth is I liked it better than bacon! I would let the oil linger in my mouth and savour the flavour. I'm not saying I was an odd child — at least I hope I wasn't — but I truly enjoyed the taste of what I later learned had the benefit of being anti-inflammatory. And today, I am just as enthusiastic about supplementing my diet with nutrients to stay healthy and reduce the effects of ageing, from disease to wrinkles to weight gain.

I think nature's pharmacy should always be the first place we turn for wellness, as throughout history plants have been shown to promote vitality and restore health. Modern science is born out of this theory, showing that plants can be used to prevent disease and promote healing and wellness.

That's the reason I was, and still am, so passionate about discovering the ability of plant extracts to activate autophagy. The first botanical I learned about was bergamot, the fruit Dr Janda touted for its youth-boosting benefits (see page 65). Since then, Dr Janda has become an inspiration in the development of Glow15. She recently shared an article published in the *Journal of Pharma Nutrition* citing her findings that protein activation in association with upregulation of autophagy can be achieved by supplementing with bergamot. Dr Janda was able to show that this powerful plant could help protect and repair your cells to change the way you age. She attributed this to the concentration of polyphenols in the pith of the fruit.

Dr Janda helped me discover that polyphenols – naturally occurring compounds that give plants colour and protect them from disease and other environmental threats – are the most efficient and effective nutritional supplements to defy cellular ageing. As you may remember from the Glow15 diet guidelines, autophagy-activating polyphenols like red wine, cocoa, berries, nuts, seeds and leafy greens can promote your well-being; but I want to introduce you to the mightiest superhero of all polyphenols: Powerphenols.

Pow·er·phe·nol
pou´ər-fē´nôl´,-nōl´,-nŏl´ /*(noun)* an autophagy-activating polyphenol in a nutrient-dense form *with an ability to defy ageing and improve health*

I coined the word *Powerphenol* because I've found that a high potency of the active nutrient polyphenol is one of the best ways to take years off your age.

These are specific, potent antioxidants that not only protect your cells but also activate autophagy to repair them, for the ultimate youth-boosting benefits. It's a one-two power punch – protect and repair, as both of these functions can reduce the risk of diseases like diabetes, heart disease, cancer and neurodegenerative disorders. In addition, this unique combination of antioxidant plus autophagy activator has been shown to optimize metabolism so you maintain high energy levels using fewer calories, increase beneficial bacteria in your gut, boost longevity and improve your overall health and appearance.

In this chapter, I'll reveal my Powerphenols and how you can take full advantage of their youth-restoring value.

The Powerphenol Purpose

The "upgrade" is a powerful tool in science and tech — and, really, all of life. Just think about all the ways that performance and experiences can be improved with power-packed add-ons: you can add memory to a computer, horsepower to an engine, data to a phone plan, legroom to an airline ticket. These bonuses are designed to give you more than what you can get from the basics. That's exactly what Powerphenols do. They upgrade your health and wellness by acting as a nutritional advantage — allowing you to not just survive, but thrive. Here are four reasons you should incorporate Powerphenols into your life.

1. Food Production Is Flawed

Modern food production practices optimize for quantity, not quality, which can lead to nutritional deficiencies. Water and soil are both necessary for plants to grow, but both have been degraded over time. Soil is increasingly depleted of essential nutrients. And water is also robbed of its nutrient value through modern filtration, draining it of important minerals like magnesium, according to the *Journal of Environmental Health Perspectives*. In addition, it is much more likely for water today to be contaminated with potentially harmful chemicals. All this can have an impact on your food, potentially leaving you lacking important nutrients.

2. Nutrient Levels Are Dropping

The *Journal of the American College of Nutrition* revealed nutritional data from 1950 and 1999 for forty-three different vegetables and fruits, and found that the amount of protein, calcium, phosphorus, iron, riboflavin, and vitamin C they contained had declined over the past fifty years. This is not an isolated study, as research shows nutrient levels have dropped over the past two decades. According to *Scientific American,* you would have to eat eight oranges today to derive the same amount of vitamin A that your grandparents would have got from one orange. Scientists speculate that one reason for this decline is pesticides.

For example, higher amounts of polyphenols are found in organic and sustainably grown fruits and vegetables. This is because plants produce polyphenols to protect themselves against insects and viruses. When pesticides are used, there is no reason for plants to naturally defend themselves. So even if you follow a healthy diet and eat fruits and vegetables, you can't always rely on the quality of your food. This may put you at risk for nutrient deficiencies that can damage DNA and make you age faster.

3. Absorption Declines with Age

It's a biological fact that, as you get older, it becomes more difficult for your body to break down and absorb nutrients from food. There are a variety of reasons for this, including a reduction in saliva, which helps break down food, a decrease in gastrointestinal enzymes, and hormonal changes. As a result of this malabsorption, we need more nutrients as we age, from the most bioavailable sources possible.

4. Your Nutritional Insurance Policy

While getting nutrients from food should always be your focus, it is not always realistic. Soil and water depletion, pesticides and Accelerated Agers like environmental toxins, lack of sleep, and poor absorption can all cause nutrient deficiencies. Relying on food alone puts your body and brain at an unnecessary disadvantage − making you susceptible to age-related disease.

A study published in the *Journal of the International Society of Sports Nutrition* titled "Food Alone May Not Provide Sufficient Macronutrients for Preventing Deficiency" showed that every single one of the seventy athletes who participated in the study was deficient in more than two nutrients.

But nutritional supplements like Powerphenols can give you a great advantage and work as an insurance policy to protect you and help prolong your life.

Your Shield of Wellness

In nature, polyphenols have a clearly defined role: protecting plants against aggressors like UV rays, insects, and disease. And when you ingest these botanical allies by taking Powerphenol nutritional supplements, you not only get similar protective benefits, you also create a shield of wellness that both defends and fights against the visible and invisible signs of ageing. Think of them as the botanicals you need in the battle for beauty, youth and vibrancy.

This wellness shield defends your cells by preventing cell oxidation and fights cellular ageing. Potent polyphenols have been shown to defend against long-term complications from diabetes, cardiovascular disease, cancers and neurodegenerative diseases.

I believe these nutritional superstars are the ideal weapons in the battle to boost our youth. And I've identified four Powerphenols to help you look and feel younger: resveratrol-trans, organic curcumin, berberine and EGCG.

Youth-Restoring Powerphenols

Powerphenols can give you a biological youth boost when used to supplement your Glow15 diet. Take these nutrient-dense superstars on High days, as they can produce the same effects on metabolism as fasting. Powerphenols are fat soluble, meaning that to ensure absorption, you should take them with fat. Try them with an AvocaGlow (page 200), Egg15 (pages 201–16), or AutophaTea (page 199) in the morning. These fats will help your body absorb more polyphenols.

And while the higher levels of the concentrated active ingredients in Powerphenol supplements can boost protection and repair your cells, you should check with your doctor before taking them if you are pregnant or taking prescription medication.

Resveratrol-Trans

Enjoy a glass of red wine. Indulge in some dark chocolate. Eat a tablespoon or two of peanut butter. They contain a Powerphenol called resveratrol.

Yes, that same resveratrol I discovered during dinner with my cousin in France. It was then that we spoke of the French paradox, the relationship between the amount of wine commonly consumed and the evidence of a lowered risk of heart disease among the French. The explanation for this phenomenon, according to scientists: resveratrol, a Powerphenol found in grape skins and other plant sources, including peanuts and berries.

There are thousands of studies showing that resveratrol helps slow the ageing process. It does this by promoting longevity via a class of proteins called sirtuins. I had the opportunity to spend time with and learn from Dr David Sinclair, the foremost researcher on resveratrol, who discovered that it was the most powerful sirtuin 1 activator of all the natural compounds he tested.

Resveratrol-trans is the best type of resveratrol to enhance autophagy. It has been clinically proven to provide anti-ageing benefits. The other type of resveratrol, called resveratrol-cis, has been shown to be much less stable and effective than resveratrol-trans. The benefits of resveratrol-trans have been linked to nearly every category of anti-ageing.

Younger Brain: Resveratrol-trans has been shown to increase cerebral blood flow. One Georgetown University study found that resveratrol-trans can help by reducing molecules that enter the brain and can contribute to cognitive decline; it does this by restoring the integrity of the blood-brain barrier. The study showed that resveratrol-trans worked by reducing inflammation in the brain, thus slowing down the decline in brain function among patients. This was the largest nationwide study done on people with mild to moderate Alzheimer's disease testing pure high-dose resveratrol-trans.

Healthy Weight: Resveratrol enhances mitochondrial function – through sirtuin activation – which can improve metabolism. Also, resveratrol has been shown to reduce body weight and fat mass by helping to control appetite. In addition, it helps reduce the proliferation of fat cells and decrease the production of stored fat. A review of studies in the journal *Annals of the New York Academy of Sciences* found that the anti-obesity effects can come from various mechanisms, including the ability of resveratrol to prevent the development of insulin resistance.

Anti-Inflammatory: Resveratrol's role in insulin resistance has been shown to reduce blood glucose in patients with type-2 diabetes. Evidence even suggests that resveratrol-trans is a more potent anti-inflammatory agent than NSAIDs such as aspirin and ibuprofen.

Healthy Heart: This Powerphenol has been shown to improve cardiovascular health by supporting arterial health. It's also been shown to improve circulation and blood flow.

How to Take It: You would have to drink hundreds of glasses of red wine to get the levels of resveratrol-trans that are used in studies, so it is best taken daily as a nutritional supplement. This is why it became one of my favourite anti-ageing ingredients. I sourced the highest-quality organic French red wine grapes and organic Muscadine grapes to create a superior supplement that's been used in studies around the world, including universities and the National Institute of Health (NIH), and millions of people have benefited from the power of resveratrol-trans. It's safe to take up to 1,000 mg of this incredible Powerphenol daily.

Organic Curcumin

When I was travelling in India, I visited Bangalore, considered the country's Silicon Valley, to meet with scientists and farmers to learn more about curcumin – the bright yellow phytonutrient found in the turmeric plant, grown throughout Asia. Everywhere I went, I experienced the fragrance of the spice. At the farmers' market, everyone was buying turmeric. When I asked the women shopping how they used the plant, they told me they did the same things as their mothers and their grandmothers did: grated it, cooked with it and made sure to add at least a teaspoon a day to their diet. The reasons they gave me? It wards off bacteria, prevents sickness and alleviates sadness. While they may sound like folklore, medical science supports these women's beliefs. That's why turmeric's most active ingredient — the yellow-coloured chemical called curcumin — is one of my favourite Powerphenols.

You likely know or have heard of curcumin because turmeric is the main spice in many curry blends. Its health benefits were first published in scientific findings in the early 1800s, and now modern science has discovered that it can induce autophagy.

Reduce Inflammation: Studies show that it may be comparable to some popular anti-inflammatory drugs, like NSAIDs.

Boost Mood: This Powerphenol also works as a natural anti-depressant — and studies show it helps quell anxiety, too. A study in the journal *Phytotherapy Research* showed that curcumin was as effective as Prozac in alleviating symptoms of depression.

Reduce Blood Sugar: High levels of blood sugar are bad for all kinds of reasons; they lead to fat storage, put you at risk of developing diabetes and increase your risk of arterial damage and heart disease. High glucose levels mean your body will age faster biologically, not to mention the fact that they can also affect your appearance. New research shows that people with type-2 diabetes who had elevated blood sugar were more likely to "look" older than those with lower blood sugar, who look "younger" than their age. The good news: curcumin helps to lower blood sugar.

You can amp up your fight against high blood sugar with curcumin, which helps regulate and improve insulin's ability to bind to sugar. It's also been shown to reduce the activity of specific liver enzymes that release sugar into the bloodstream while also activating enzymes that store sugar as glycogen. Note: Because curcumin can lower blood sugar, anyone taking diabetes drugs or insulin will want to monitor their blood sugar levels while taking it as a nutritional supplement.

Promote Heart Health: Research shows that turmeric or curcumin supplements can reduce total cholesterol, as well as LDL ("bad") cholesterol and triglycerides.

How to Take It: Research shows that taking 500 mg twice a day is best. Take organic curcumin that has been cleanly extracted using carbon dioxide. Organic is important, because curcumin is extracted from the turmeric root, and since the active ingredient has been condensed, you want to make sure you are not getting any herbicides or pesticides.

I always take curcumin with a fat-first meal, which dramatically boosts absorption of this fat-soluble Powerphenol. You can also take it with black pepper or a piperine supplement to increase the absorption.

Berberine

How many times have you wished you could pop a pill instead of working out? Well, while berberine can't replace a good sweat (but don't despair – my Glow15 exercise programme, which you'll read about in the next chapter, will be pretty easy to swallow!), this Powerphenol has been called "exercise in a pill". Berberine's superpower is its ability to produce biological effects similar to those of exercise.

Grown throughout Asia, berberine has been used for centuries as an alternative medicine. It's difficult to find in food, which is why I get it in a nutritional supplement. In studies, berberine has been shown to benefit diabetes, metabolic syndrome, and gut health, but I am most interested in its relationship to fat.

Now, it's important to know that we all have distinct types of fat in our bodies.

White fat is the main form of fat found all around your body – your belly, thighs, arms, hips and everywhere else. An excess of white fat stored deep around your organs, called visceral fat, can be associated with diabetes and heart disease.

The second type is brown fat. Brown fat is actually beneficial. It can help your body burn, not store, calories. Your brown fat is normally stored in your upper back and neck. Studies show that leaner adults have more brown fat than those who are overweight, but since our bodies don't have very much to begin with, the focus is on trying to figure out ways to generate new brown fat.

So what does berberine have to do with it all?

Burn Fat: Berberine enhances activity in brown fat – by helping to generate the heat that burns more calories. Clinical trials have shown that berberine is one of the few compounds that can activate autophagy through a protein called AMPK. This protein is so powerful that it's often referred to as the "metabolic master switch", as it mimics the heat-activating process that burns calories (thermogenesis). Recent studies on mice show that berberine helps the body deal with weight gain and disorders associated with weight gain. And other studies show that berberine can help manage genes to have a positive effect on burning calories from fat and keeping the body from creating fat in the first place, as well as fighting insulin resistance, a contributor to weight gain.

Another study showed that people who took berberine three times a day lost an average of 5 pounds. While that may not seem like substantial weight loss, if you consider that it was the only variable measured, it is an extra boost to combat fat!

If its role in fat burning is not enough, this superhero also has properties that make it a wonderful addition to your Glow15 Powerphenol arsenal.

Boost Gut Health: Berberine has been shown to have a positive effect on your gut flora. Improving the bacteria in your gut can not only ease digestive issues, but also can play an important role in your entire body. We know that "beneficial" or "good" bacteria do so much more than protect us from bloating, gas and traditional GI issues. These healthy bacteria might be the commanders in chief of your well-being and ageing.

The better you are able to maintain a high ratio of healthy to harmful bacteria, the stronger and more vibrant your overall health and wellness will be. Berberine can help eradicate bacteria and optimize your gut microbe status. So if you want to feel well and look your best, look after your gut health first.

Reduce Blood Sugar and Cholesterol: A study published in the journal *Metabolism* showed that berberine worked to manage blood sugar and lipid metabolism as effectively as the prescription drug metformin. Here's an interesting fact: berberine reduces blood sugar *only* if blood sugar is elevated. Berberine has also been shown to reduce cholesterol and triglyceride levels without the side effects of statin therapy, the conventional pharmaceutical treatment for high cholesterol.

Improve Lung Health: The journal *Inflammation* published a study showing that pretreatment with berberine can reduce inflammation in the lungs and also help to reduce injury related to cigarette smoking.

How to Take It: I suggest taking three 500-milligram doses a day to keep your blood levels stable. I also suggest taking each dose with a meal to help avoid stomach upset.

EGCG

This Powerphenol is found in green tea and is a key ingredient in my AutophaTea (see page 199).

EGCG (epigallocatechin gallate) is a compound made from two others, and it's classified as a catechin – a kind of flavanol shown to be beneficial to your body. It's found in high amounts in green tea; in fact, EGCG makes up 50 to 60 per cent of green tea catechins.

A cup of AutophaTea's or EGCG extract's youth-promoting benefits have been well documented.

Boost Longevity: At the end of each strand of DNA, you have little tips called telomeres (scientists often liken them to the little plastic pieces at the ends of shoelaces), which have a big role in how well you age. When telomeres get damaged or frayed, it influences the way your genes function. Over time, your telomeres shorten; this is a natural part of ageing. But when they get too short, that's when your DNA is more at risk. EGCG can protect your telomeres from damage, thus increasing your youth span and reducing the risk of developing disease. The *Journal of Experimental Biology* links the EGCG in green tea to a healthier life and boosted longevity.

Lose Weight: This Powerphenol can suppress appetite, reduce fat and burn calories. In addition, it can also help boost metabolism. A study from Geneva indicated that consuming EGCG with caffeine can increase metabolism by 4 per cent, which means you can burn calories after taking it, even if you aren't expending a lot of energy.

Fight Cancer: As an antioxidant, EGCG can fight free radical damage that could lead to cancer. One study in the journal *Cancer Metastasis Reviews* explained that EGCG's power to reduce cancer risk is twofold: One, the substance was shown to help inhibit the growth of tumors. Two,

GLOW15 SUCCESS STORY

"I'm no longer frightened, I'm motivated" KRISSY

"**What I like best about Glow15 is how it works with my body, making me feel better and healthier.** I'm a registered cardiovascular invasive specialist. I spend most of my days scrubbing in with physicians after a patient has a heart attack. The vast majority of these patients are overweight. So am I. I worry that it could be me on that table next. I have high cholesterol and what doctors call a fatty liver and that combined with my weight means I should be scared. But that fear can be paralysing. Glow15 is helping me move forward and claim control of my own health. I know I have a long way to go, but **I've already lost 3 pounds and some inches, and feel more awake and alert.** I made changes to the way I eat and sleep and move, but the biggest change for me was taking Powerphenols. I tried the resveratrol and the curcumin. And while I can't attribute the changes I see and feel to any one part of the plan, I do feel like those nutritional supplements gave me a little extra boost or helped jump start my autophagy. Most importantly, I believe they are working to help make me healthier. I'm actually looking forward to going back to my doctor and getting my cholesterol levels checked. **The best part of Glow15: I'm no longer frightened, I'm motivated.**"

it was also shown to help decrease the factors that lead to metastasis – the spreading of cancer to surrounding tissues (often the factor responsible for cancer deaths).

Improve Brain Function: According to the journal *Appetite*, EGCG can strengthen and calm your brain. In the *Journal of Nutritional Biochemistry*, the Powerphenol EGCG is now being considered as a therapy to alter brain ageing processes and protect against neurodegenerative disorders like Parkinson's and Alzheimer's.

How to Take It: I think the best way to get this Powerphenol is through AutophaTea, which has 70 to 150 milligrams of EGCG per tea bag. You can also find it in capsule form. It is safe for most adults to have up to 600 milligrams of EGCG daily. It is best to spread out your intake over three increments and take it with meals to aid absorption.

Think of Powerphenols as a dual dose – antioxidant and autophagy activating – of plant medicine to help you grow younger. They protect and defend. They prevent and repair. And, they can help your cells act younger, giving you the ability to change the way you age. I believe Powerphenols can be your wellness shield, revitalizing your body to fight disease, promote weight loss, and boost energy.

Want to further those benefits? In the next chapter, I'll show you how a minimum amount of exercise can lead to maximum youth. Get ready. Get set. Let's glow!

How to Know If Your Herbal Supplements Are High Quality and Safe

Buy from a reliable source. Most supplements sold in the UK are regulated by food law. To ensure your supplement is made to the highest standards and contains the ingredients listed in the quantities stated, buy from a reputable source and check the company adheres to GMP (good manufacturing practice).

Look for the THR mark on herbals. Some herbal products in the UK are sold as licensed medicines, which means they comply with pharmaceutical standards relating to safety and manufacturing. If you can find a product with the THR (traditional herbal remedy) label, go with this.

The back up is to make sure the manufacturer uses GMP and to look for "standardized extract" somewhere on the label – this should mean the herb potency is uniform batch to batch.

Choose single-ingredient supplements. They're more likely to contain the active ingredient listed on the label and are less likely to be contaminated.

Chapter 6

Ready, Set, Glow!

Minimum Exercise for Maximum Youth

Exercise — just the word can put me on the defensive. And I know I'm not alone. We all have our go-to excuses for why we can't work out today.

I'm too tired. This is the most common reason most people don't exercise. Know this, though: Research reveals that working out can actually make you more energetic.

I don't have time. You're busy, of course. But the time is there. There are 24 hours in a day and 7 days in a week. So if you aim to spend 8 hours a day sleeping, and if you have to use about 40 to 50 hours a week working — you still have between 60 and 70 hours left! On Glow15, you only need 2 of those hours — total — to reap the benefits of exercise. So what are you going to do with all your free time?

Exercise makes me eat more. Actually, the opposite may be true. A study at Brigham Young University in Utah found that exercising may actually suppress your appetite.

We've all come up with reasons why we can't exercise — but the truth is, exercise is your best defense against ageing. It helps you build bone, gain muscle and lose weight — and it can boost your mood.

But did you know exercise can also boost your autophagy and add years to your life? In this chapter, I'll give you the Glow15 excuse-proof strategies – and help you change your exercise psyche from a defensive response to an offensive one.

How Exercise Activates Autophagy

Fear of exercise may actually be good for your autophagy – that is, if it can elicit acute or healthy stress. Remember, stress is good for promoting autophagy. Autophagy loves stress! And creating healthy stress can help you get the most out of your workout so you look and feel younger.

There are two common types of stress: acute and chronic. Acute stress is temporary. It can be caused by one-off triggers, like a surprising or embarrassing moment or hitting your funny bone. Chronic stress is long-lasting – like financial worries or arthritis.

One of the biggest differences between the two is their impact on our health. Acute stress can often be healthy – it can help us move faster during an emergency or study harder before an exam – whereas chronic stress has been linked to health problems, including heart disease, depression and obesity.

We've all heard the stories of the fight-or-flight response. Think of the adrenaline-driven mother who found incredible superpowers to save her child by lifting a car or fending off an attacker. The cause here is acute stress and the response is physical activity. And that does more than just reunite mother and child. That stress response boosts autophagy to create more beneficial cell components and remove negative ones.

When we exercise, we're actually putting our bodies under acute stress. When we go for a run or a bike ride or climb up a flight of stairs, our heart rate accelerates to pump blood and oxygen to our muscles. Our hearts don't want to beat that fast all the time, but they need to do so to help us get through the session.

That's a form of stress. The same thing happens when we lift weights or do any other form of exercise. The actions we take when we work our bodies cause microscopic tears in our muscles. That sets autophagy in motion, as you're now in stress-response mode – and autophagy activates to remove and repair the damage. This makes your muscles stronger and more resilient – in anticipation that you may just do that same thing again and your muscles need to be better prepared for the stress the next time it happens. The result of an exercise session? Short-term stress for long-lasting benefits.

Until recently, few researchers looked into the effect of exercise on autophagy within cells. Once they did start studying it, they observed that exercise seems to have the ability to boost autophagy, specifically in cells found in bones and the heart. Autophagy rates increase when cells are starved or when they're placed under physiological stress, like increasing heart rate and straining muscles. That's exactly what happens during exercise. In fact, it's such an exciting part of research that some have suggested that the true benefit of exercise is that it helps get rid of our toxic waste via autophagy, and that's what provides all the long-term benefits like improved cardiovascular health and immune system.

One of the leading pioneers in autophagy research, Dr Beth Levine at the University of Texas Southwestern Medical Center, showed that exercise activates autophagy in muscle cells. She did this by looking at mice running on treadmills – first seeing autophagy increase at 30 minutes, then maximizing to 100 per cent at 80 minutes, and then plateauing. The American Aging Association recently ran its own human study in older adults, which supported Dr Levine's findings.

A related benefit? Exercise helps burn fat. Because excess body fat gunks up the autophagy process, exercise is like a one-two autophagy punch – it improves the system by stressing the cells *and* by getting fat out of the way. Dr Heather Hausenblas, associate dean of the School of Applied Health Sciences at Jacksonville University in Florida, says that exercise is like the "high-speed Internet" compared to the "dial-up" in the cellular cleanup process. That's because exercise cleans up the waste inside the cells very quickly (the high-speed Internet); it moves very slowly when we just sit around (dial-up).

The best part: You can expect instant results. Autophagy ramps up during and immediately after one single bout of endurance exercise. And it has long-lasting results. Researchers in Austria found that older people who regularly worked out throughout their adult lives were stronger and more youthful. That manifests itself not only in things like bones and muscles, but also at the microscopic cellular levels—in that "older" bodies present themselves to be much younger than they actually are. So no matter your current age, exercise can slow the age-related decline in autophagy and make a significant difference in offsetting muscle atrophy and extending longevity.

Are you ready? Are you set? Let's glow!

The Glow15 Workout Guidelines

I don't want you to think of exercise as another sacrifice you have to make. Instead I want you to reframe the idea of exercise and think of the Glow15 workouts as a great way to make yourself a priority.

Instead of thinking that exercising is just another thing you have to do, think of it as an opportunity to take back your time. Do it for yourself. It's a simple way to be bold. Have the courage to make an excuse to make time for you. This is not selfish; this is self-health.

And when you do make that time for yourself, you will find that you are so much more present in your life – from work to family. Taking time for yourself actually allows you to give time to others. Practicing self-health allows you to be more focused, calmer, and more grounded. And in turn, this will help you be a better mother, wife, friend, worker, confidante, caregiver – woman!

I know that the first step can be the hardest – whether you're exercising for the first time or making a change in the type of workouts you do. Remember, fear of change is usually harder than the change itself. But if you can get out of your own head, I promise you that the results can be transforming. Not just physically, but mentally, too. The high that you get from accomplishing these workouts and the strength that you will feel, I believe, will become two of your most valuable assets.

GLOW15 SUCCESS STORY

"Exercise helped get me off my meds" SHARON

"Glow15 not only transformed my health, it changed the way I feel about myself. It even inspired me to help others make the same changes. Before this, I was overweight, pre-diabetic and taking medication for high blood pressure. I knew I was setting a bad example for my daughter. The ironic part: I am an executive operations coordinator for medical professionals. I spend my days helping others better their health, but I was neglecting my own. Then, I started Glow15 and began paying attention. I changed my diet and my sleep schedule, but the biggest change I made was exercise. I had never really worked out before, and while at first it was really difficult, I kept at it. And I kept keeping at it, huffing and puffing my way through group fitness classes until it became fun. **After just 15 days of doing HIIT and resistance training, I lost 6.5 pounds!** I started to look forward to class, and I recruited my coworkers to join me. We continued to see changes – not just in our weight, but in our attitudes. Working out made us all happier and more positive, and that made us keep moving. **By the next fifteen-day cycle I lost over 13 pounds and reduced my body fat by 5 per cent.** And I'm proud to say **I no longer need my blood pressure medication.** And my daughter took notice and started to follow my example. Together, we are committed to making Glow15 a permanent part of our lifestyle."

Here's the best part about Glow15 exercise: less is more. You don't actually need a lot of time to get results. You can be smart and efficient – boosting your autophagy quickly and effectively by following some specific guidelines.

Exercise on High days only. Remember, High days are when you're not fasting and are eating normal to high amounts of protein. So that means you will *not* exercise every day. Again, the most natural and efficient way to activate autophagy is through stress. And the two best ways to do that are fasting and exercise. You don't need to exercise on fasting days because it's important to alternate how and when you stress your cells. As you already know, autophagy is best optimized when it's turned on and off, like breathing in and out, or in the case of exercise, huffing and puffing. Working out these four days a week is what allows our bodies to flip that switch. It's a myth that we have to exercise for hours on end to achieve the best results. You will see results in two weeks only working out eight total times, and – especially if you haven't done much exercise recently – you will feel different instantly. Stronger, younger, more energetic and, yes, more sore. The mood-boosting benefits extend to hours (and often days!) beyond your workout.

All you need is 30 minutes. That's right – just 30 minutes to boost your autophagy and get anti-ageing benefits for your brain and body. According to Dr Beth Levine, exercise can be even faster than nutrient deprivation at inducing autophagy. In her experiment, she proved that after just 30 minutes on a treadmill, autophagy is induced, and further concluded that exercising at a high intensity for that amount of time further activates autophagy by 40 to 50 per cent. It is important to note that the rate increased up until 80 minutes on the treadmill, at which point it plateaued. In the journal *Nature,* Dr Levine says that "autophagy may represent a cellular mechanism by which exercise prolongs life and protects against cancer, cardiovascular disorders, and inflammatory diseases". Her findings not only gave her an excuse to not exercise for a lengthy amount of time, but inspired her to get her own treadmill.

Run on empty. Exercising on an empty stomach can increase how much autophagy your cells engage in, indicating again that the stress of exercise takes a greater – and quicker – toll on our bodies when we're not adequately fueled.

Your body needs more nutrients during exercise than during rest, and this makes your cellular cleansing cleanup crew work harder. So try exercising first thing in the morning if you can. If that's not possible, you should still exercise on an empty stomach; just wait 1 to 2 hours after eating before you work out.

Pregame with caffeine. Instead of eating, drink caffeine before your workout. Not only does caffeine promote autophagy, it can actually help you exercise better and boost the fat-burning benefits. According to a new study published in the *Journal of Applied Physiology*, drinking caffeine – like my favourite AutophaTea (see page 199) – prior to exercise not only gives you additional energy, it also improves your performance. And the boost from caffeine may help you burn fat.

Turn up the heat. Another way to increase autophagy? Heat shock. A study in the journal *Autophagy* explains that heat stress – achieved by exercising in temperatures above 30°C to raise your core body temperature – boosts autophagy by activating genes that optimize heat shock proteins inside your cells. These heat shock proteins help prevent plaque formation in your brain and vascular system and are also involved in longevity. While it may seem counterintuitive when you are already planning to sweat, added heat can actually be good for you. There are a few ways you can initiate heat shock when exercising, like turning up the room temperature or wearing extra layers or working out outdoors on a hot day. You can also raise your core temperature immediately after exercise to get the benefits by taking a hot bath or using a steam room or sauna. Note: be careful to not overdo it in the heat, and make sure you drink plenty of water.

Refuel with protein. Remember, you're working out on a High day, which also means you're eating normal to high amounts of protein. After your workout, make sure you refuel with protein. This may seem counterintuitive, because high protein consumption turns off autophagy. But studies show that consuming just one serving of protein once you're through with a workout (especially one that involves strength training) raises the level of autophagy occurring in your muscle cells. Researchers say this is all part of the adaptation your muscles undergo when you start taxing them with resistance exercises. So aim for one serving of protein (try one of the High day Egg15 recipes on pages 201–16!) within 30 minutes after cooling down from your workout.

Keep on going. Don't give up. Dr Hausenblas warns against taking too much time off. She found a study that showed the autophagy benefits of exercise disappeared if people stopped working out for two weeks. So let these fifteen days serve as an inspirational foundation to keep on going. Dr. Hausenblas also suggested that you shouldn't be afraid to get out of your comfort zone and try something new – make sure you keep adding challenges to your workouts to give your body new (healthy) stressors that trigger autophagy. There's preliminary data to suggest that the novelty of a new exercise initiates more autophagy as well. That's because the more you challenge your muscles in novel ways, the more opportunities you offer them to adapt and make use of your cellular cleansing cleanup crew.

Before you keep going, you need to get started. And I know that first step can be the hardest one. One of the best ways to get past your fear–of exercise, of a new routine, of change – is to properly prepare. Here, that means figuring out what works best for you: are you motivated by music? If so, download a new playlist. Will you be more accountable if you enlist a workout partner? If so, get a friend to "glow" with you. Do you find the gym intimidating, too far to get to, or are you just not ready to let others see you sweat? If so, my youth-boosting Glow15 workouts (see pages 264–78) can be done in the comfort of your own home. The point is that you should think about what backdrop – whether it's environmental or interpersonal – will help kindle your motivational fire. I'm confident that once you get into a groove, your fire will be fully stoked. That's because once you see and feel the effects, you will want to experience more.

Above all, remember, the point is not perfection. Perfect is boring. Perfect is basic. Perfect is pointless. The best thing you can do for your body and brain is to challenge them. This will activate autophagy and keep you engaged, motivated, and looking and feeling young!

Glow15's youth-boosting workouts can be adapted to any fitness level – whether you're a novice or a more advanced athlete. But this doesn't mean you need to join a gym or buy fancy exercise equipment. Instead, they are designed to work with your lifestyle to most effectively boost your autophagy in the least amount of time.

The Glow15 Workouts

There are two types of exercise scientifically proven to best initiate autophagy. They will allow you to work and challenge your body in different ways. And in just fifteen days, you will notice a difference. In the way you look. In the way you feel. In what you can physically do.

You'll do two high-intensity interval training (HIIT) workouts and two resistance exercise training (RET) workouts each week.

High-Intensity Interval Training (HIIT): Don't be alarmed by the words *high intensity*. HIIT involves working in short bursts until you're out of breath. This is beneficial because that limited deficiency of oxygen activates autophagy. HIIT workouts pair low-effort periods with high-effort periods: work hard in short bursts and then have time to recover before doing it all over again. So, for example, you might do 1 minute of harder effort, followed by 1 minute of moderate effort — in any form of exercise you like. Because intensity is measured by your heart rate, the exercise is customized for you. A beginner, for instance, may find a brisk walk or jog will be their high intensity, whereas someone who works out regularly may need to sprint to reach their target heart rate. Best of all, the total workout time is short. You need to do just 30 minutes to see the benefits.

Now, you may be asking, What exactly are moderate and hard effort? Good question, and this depends a bit on your fitness level, since there's no way to standardize effort to every woman. How do you determine what high intensity means for you? One of the best ways is to figure out your target heart rate. According to the American Heart Association, to find your target heart rate, you have to calculate your resting heart rate, or the beats per minute when your heart is at rest. I've given you instructions for this on page 155. A resting heart rate under 80 beats per minute is considered healthy. The more you exercise, the lower your heart rate tends to be.

Next, calculate your maximum heart rate — this is the maximum number of beats per minute your heart can handle during exercise. The basic way to do this is to subtract your age from 220. For example, if you're 40 years old, subtract 40 from 220 to get a maximum heart rate of 180.

And finally, your target heart rate or your anaerobic threshold – the ideal level at which your heart is being worked but not overworked – for the Glow15 HIIT exercises is between 70 and 85 per cent of your maximum heart rate.

So if you're 40, and your maximum heart rate is 180, you multiply 180 by .7 to get 70 per cent of your maximum, which is 126. So your target heart rate would be 126 beats per minute. For a higher target – 85 per cent of your max – you would aim for 153 beats per minute. If you are a beginner, it is best to start with a target heart rate of 70 per cent and work your way up to 85 per cent. This obviously is easier to measure with a heart rate monitor or activity tracking device, which you can buy online. I always use one.

Another (somewhat less mathematical and easier) way to determine your high-intensity level is through perceived exertion scale. Think of effort as being on a scale of 1 to 10, with 10 being all-out effort, like a sprint, and 1 being a leisurely walk – sitting on the couch would be a 0. Moderate effort would be a 3 or 4 on the scale and harder effort would be in the 6 to 8 range. Now imagine how that works in practical terms. Let's say you go out for a run. You would jog easily (an effort of 3 or 4) for a minute and then follow that with a minute when you push the pace and are huffing and puffing (that's your 6 to 8 on the scale). Alternating between the two for a period of time (in this case, 30 minutes) is how high-intensity interval training works.

What's great about this kind of workout is that you can do it in a variety of forms, such as running, swimming, spinning, rowing, etc. . . . Plus, some sports are naturally high-interval, like tennis, when you're going back and forth between hard effort and then periods of rest in between points. There are also many popular fitness classes that use HIIT – such as Barry's Boot Camp, Virgin Active The Grid classes and Les Mills Sprint and other HIIT spinning classes combine periods of intense activity with moderate exercise.

The reason this works so well is that our bodies cycle through anaerobic exercise in which we're deprived of oxygen; the scientific term for this is hypoxia. When you feel out of breath after pushing yourself, that's your body trying to get oxygen. Now, that doesn't exactly seem like it would be a good thing, but the benefit is that the deficiency of oxygen in short bursts activates autophagy. And with HIIT, this happens when you reach your anaerobic threshold – your target heart rate or the high end of your perceived exertion scale.

A study in the journal *Cell Metabolism* found that interval training improved the health and number of the mitochondria in cells. This is important because mitochondria serve as the power plants of the cells, meaning that improvement in their health can translate to more energy. In this study, older people experienced nearly a 70 per cent increase in the capacity of the mitochondria, while others had a nearly 50 per cent increase. Researchers at McMasters University in Canada have done lots of work in this area, showing that interval training helps improve age- and disease-related indicators such as blood pressure and cardiovascular health. One of the leading researchers reported that one of the best parts about interval training is that you don't have to reach a certain level to experience benefits – it's all relative to the individual. "You just need to feel some brief discomfort," he said. And that's a good way to put it; those periods of higher effort on the scale won't feel all that comfortable, but they're momentary. And that momentary stress is what creates the autophagic effect.

You should do HIIT two days a week, and again, I want you to pick whatever activity you most like. (More details about how to exactly do this workout are on page 264.)

Resistance Exercise Training (RET): What gives your body tone and that lean and strong look? Well, that's resistance training. This workout is about strength training, where you push against or fight a force in order to exercise your muscles. RET activates autophagy by stressing cells in your muscles; this is key to maintaining muscle mass, which decreases with age.

You might immediately think of lifting weights as a form of RET, and you're right, but that doesn't have to only mean lifting heavy weights. You can practice RET with your own body weight – push-ups, lunges and squats are all forms of very good resistance training. Yoga can be a form of RET. Or you can even use normal household items, like bottles of water, bags of rice, books or even household supplies like laundry detergent.

The point of this kind of workout — which, again, you will do two times a week — is that you will be putting stress on your cells to eventually rebuild muscle tissue and thus activate your autophagy.

RET remains the most effective treatment for the loss of muscle mass and strength. Evidence suggests that autophagic signalling is altered in aged skeletal muscles. But a study in the journal *Experimental Gerontology* found regular resistance training activates autophagy and helps prevent the loss of muscle mass while improving muscle strength. One study from the University of New Mexico even showed that RET changed the way that genes were expressed in an older group of people, and the researchers noted that it was an indication of not only slowing the ageing process, but reversing it.

RET helps you build that lean muscle tissue, which will improve your posture and make your body better able to withstand injuries, not to mention make you look leaner and change your body shape. Perhaps even more important, adding lean tissue will actually help you burn fat. That's because muscle is what scientists would call "metabolically expensive" — that is, your muscles need a lot of calories to sustain themselves. Because your muscles need those calories, there's less chance that extra calories will be deposited and stored in your body as fat. So adding this lean tissue improves your ability to not only lose weight but also keep it off.

Don't worry — you're not going to beef up like a bodybuilder by doing RET. These workouts will help build lean muscle and help keep you fit — with strong and sexy muscles, not big and bulky ones.

The key for both HIIT and RET on Glow15 is that you can customize the workouts to your fitness level and abilities. Everything is relative. Your intensity and strength levels will increase as you get stronger and fitter, but even if you're a beginner, you can reap the benefits. Whether you get your heart rate up with a brisk walk or a full-out run or are challenged by lifting paperback books or multiple encyclopedias, if you commit to the Glow15 workouts four days a week, you will boost your autophagy and feel like a better, healthier version of you. See the specific workouts starting on page 264.

GLOW15 SUCCESS STORY

"Exercise gave me the confidence I'd been missing"

LYNDEE

"I'm a 37-year-old mum of two, and I've watched the numbers on my bathroom scales slowly increase over the past few years. Each time I get on, I tell myself that I can work off the weight. I tried all kinds of workouts and never saw the scales move. And I worked out hard! It's really disappointing to put in all that effort and not see results. I felt like a failure. But with Glow15, for the first time, I'm winning! **I not only lost weight – 7 pounds in 15 days – I also lost inches.** I tried both HIIT and RET. Surprisingly I liked the interval training most because you can vary the exercises, and even when it sucks and you feel like you can't breathe, it's over quickly, you get a break and then move on to something different. It felt like **a little bit of pain for a whole lot of gain.** And I did gain something invaluable – confidence. I stand up for myself now – and for me, that's a big deal. I guess I was a little bit of a pushover before and never shared my opinion because I didn't believe it had value. Glow15 changed that. **It gave me self-worth. That newfound backbone has made all the difference** – I love the feeling of respect I get from my friends, my family and most of all my own self-respect."

A Bedroom Exercise to Boost Autophagy

One of the most natural biological ways to create acute stress to initiate autophagy is through sex. Yes, that's right – sex can help keep you young! You will naturally turn up the heat as your core body temperature increases during sex. And not only can it be both HIIT – depending on your intensity – and RET – depending on your position (!) – but it's probably the most enjoyable way to get in your 30-minute workout! Intimate bedroom moments have given researchers new clues about combating ageing – specifically with spermidine. You may remember from chapter 4 that spermidine, which induces autophagy, is found in foods like grapefruit and cheese, but it is also an important constituent of semen. A team of European researchers, led by biologist Frank Madeo from the University of Graz in Austria, has found that spermidine is necessary for cell growth and maturation, but as cells age, their level of spermidine falls. He showed that giving an extra dose of spermidine to certain organisms and cells significantly increased their life span. Plus, if getting closer to your partner weren't enough to get you getting busy, think of the calories you can burn in the bedroom. On average, a half-hour of sex burns the same calories as a 15-minute run. According to researchers at Wilkes University in Pennsylvania, you may even be able to stave off the flu with antibodies released during sex. And Dr James Couch, a neurology professor at Southern Illinois University's Clinic, did a study on female headache sufferers. He found that nearly half of the women in the study experienced full relief after sex. And sex can help you look younger, too. An experiment at the Royal Edinburgh Hospital in Scotland found that partners who had sex at least four times per week were viewed as looking up to twelve years younger than their actual age. Hit the sheets!

Chapter 7

Glow to Sleep

Getting Quality Rest Shouldn't Be a Dream

Do you have a love/hate relationship with your bed? Love the idea of sleeping, but hate that you have so much trouble doing it? Sometimes, your bed is just a mirage. It looks so inviting, but it can feel like torture, not allowing you to sleep.

We've all been there – whether you have trouble falling asleep, staying asleep or just getting good-quality sleep, you're among the millions of adults who are sleep deprived. I was, too.

For me, it started when I was a teenager. I struggled to fall asleep at night and then had a hard time waking up in the morning for school. But at the time, and through most of my twenties, I wore my lack of sleep as a badge of honour. I told myself – and others – that I could function on fewer hours and accomplish more, and I believed that was better. But after I had my daughter, Megan, I started to crave sleep. That badge was no longer one of honour; instead, it represented defeat. I could not conquer my exhaustion.

Again, too many of us feel this way. Women actually need more sleep because our brains are more complicated, according to Loughborough University's Clinical Sleep Research Unit in Leicestershire. Plus, we are wired to suffer more than men – both emotionally and physically, according to research done at Duke University. Sleep-deprived women are more likely to suffer depression and heart disease and strokes.

Well, that's not acceptable – for any of us. I needed to find a solution for myself – and for the millions of women not getting enough quality sleep. The truth is, my search for answers started long before I discovered the power of autophagy – but once again, my cellular cleansing cleanup crew proved to be key to getting better rest. As you might expect, sleep deprivation disrupts autophagy, leading to signs of ageing. This is because autophagy is activated when you sleep. Makes sense, since we are not eating when we sleep, thereby depriving our bodies of nutrients which, as you now know, initiates autophagy. During sleep, autophagy removes the junk and repairs the damage in your cells.

So how can you get quality sleep, revive your cells and stave off disease? I consulted my team of scientists, researchers and sleep experts and devised a plan to better your sleep and activate autophagy. And then, as I did with every part of Glow15, I tested it – on myself, on a few "guinea pig" friends and, of course, on the participants in the Jacksonville University study. The results were promising, with most women reporting better sleep. But when I started speaking to the women one-on-one, they confessed that adapting the sleep plan to their lifestyle was not always easy. We don't all go to bed at the same time. We don't all need to get up at the same time. I heard and understood their frustrations. And, like them, I didn't want a one-size-fits-all approach to sleep.

And that's when I met "America's Sleep Doctor", Michael J. Breus, PhD, author of *The Power of When*. I told him about my own sleep issues and my concerns about the sleep plan designed for Glow15. Dr Breus explained that it's possible for autophagy to be out of rhythm with sleep. We each have unique cycles – and need to tailor our sleep accordingly to get back in sync with our cellular cleansing cleanup crews so that we wake rested and rejuvenated. For example, there is no statistical reason you need 8 hours of sleep per night – and Dr Breus says many women actually sleep better when they sleep for fewer hours. In fact, he referenced a study that showed that people who slept only 6.5 hours per night lived longer than those who slept more. For me, less can be more – while I used to aim for 8 hours, I now sleep a maximum of 7.5 hours per night. But I actually sleep. And I wake rested and energized. Dr Breus helped me create the Glow to Sleep plan to help you, too. This chapter will show you how to maximize your autophagy while you sleep – with a customized sleep plan for you and your lifestyle. Because when you make the most of your nights, you can make even more of your days.

Why Sleep Matters: The Autophagy Connection

You may think of sleep as time to zone out and dream, perhaps about the fun (your upcoming holiday!), the sexy (ooh-la-la), or the absurd (why is your childhood crush eating cold spaghetti in your car?). But sleep is far from just the time when your mind races with wild images and far-fetched stories. It's also the time when your body restores itself by activating autophagy. It's when your cells repair damage and eliminate toxins, proteins get folded and repaired, and neurons build new connections, all to help recharge your energy, strengthen your brain and rebuild and renew your cells.

The crucial connection between autophagy and sleep is your circadian rhythm. Circadian rhythm is your body's natural timekeeping system that regulates sleep and wakefulness – along with body temperature and hormonal changes – through one-day cycles. For most people, 24 hours is the length of a full cycle – but it can vary and is different for everyone.

Here's how it works: your circadian rhythm keeps the beat while your brain acts as a conductor. For example, your body temperature rises overnight, and your circadian rhythms should increase the tempo in the morning, helping you feel alert when you wake. But later in the evening, the beat begins to slow as your body temperature goes down, helping you relax and signalling sleep. Scientists have also identified a "second sleep" – when your body temperature drops in midafternoon. This change in the beat of your circadian rhythm is responsible for what many of us think of as an afternoon slump.

Your circadian rhythm helps maintain processes essential to your well-being, but that "orchestra" can often be out of tune. Exposure to light, work schedules or health problems, or even social commitments can all impact your cycle. Also, for most of us, as we age, our circadian rhythm naturally changes. And, most significant, that disruption of your natural cycle also disrupts your autophagy, and subsequently your body's ability to repair and renew.

So the goal of Glow15 is to synchronize your sleep and autophagy. By composing a symphony with your unique circadian rhythm, you will be better able to activate your autophagy and boost your energy in ways you've only dreamed of.

What Kind of Sleep Bird Are You?

The timing of sleep is monitored by your circadian rhythm, but not everyone keeps the same beat. Some people naturally like to stay up late, others like to go bed early, and still others fall in between. This behavioural manifestation of your circadian rhythms can be categorized into chronotypes. And your chronotype is determined by your propensity to sleep at a particular time during a 24-hour period.

Following is a popular test called the Morningness Eveningness Questionnaire. It is used to help determine your chronotype: Owl, Lark or Hummingbird.

Take the test to help determine which bird best represents you. This will help you be better able to tailor a sleep plan to your unique rhythm. Simply circle the answers that best reflect your sleep pattern.

Morningness Eveningness
Questionnaire

Adapted from Horne and Ostberg's self-assessment questionnaire, first published in the *International Journal of Chronobiology*.

1. Considering only your own "feeling best" rhythm, at what time would you get up if you were entirely free to plan your day?

   5 a.m. to 6:30 a.m. **(5)**   9:45 a.m. to 11 a.m. **(2)**

   6:30 a.m. to 7:45 a.m. **(4)**   11 a.m. to 12 p.m. **(1)**

   7:45 a.m. to 9:45 a.m. **(3)**

2. Considering only your own "feeling best" rhythm, at what time would you go to bed if you were entirely free to plan your evening?

   8 p.m. to 9 p.m. **(5)**   12:30 a.m. to 1:45 a.m. **(2)**

   9 p.m. to 10:15 p.m. **(4)**   1:45 a.m. to 3 a.m. **(1)**

   10:15 p.m. to 12:30 a.m. **(3)**

3. If there is a specific time at which you have to get up in the morning, to what extent are you dependent on being woken up by an alarm clock?

   Not at all dependent **(4)**   Fairly dependent **(2)**

   Slightly dependent **(3)**   Very dependent **(1)**

4. Assuming adequate environmental conditions, how easy do you find getting up in the mornings?

   Not at all easy **(1)**   Fairly easy **(3)**

   Not very easy **(2)**   Very easy **(4)**

5. How alert do you feel during the first half-hour after having woken in the mornings?

   Not at all alert **(1)**   Fairly alert **(3)**

   Slightly alert **(2)**   Very alert **(4)**

6. How is your appetite during the first half-hour after having woken in the mornings?

   Very poor **(1)**   Fairly good **(3)**

   Fairly poor **(2)**   Very good **(4)**

7. How tired do you feel during the first half-hour after having woken in the mornings?

   Very tired **(1)**   Fairly refreshed **(3)**

   Fairly tired **(2)**   Very refreshed **(4)**

8. When you have no commitments the next day, at what time do you go to bed compared to your usual bedtime?

   Seldom or never later **(4)**   1 to 2 hours later **(2)**

   Less than 1 hour later **(3)**   More than 2 hours later **(1)**

9. You have decided to engage in some physical exercise. A friend suggests that you do this one hour twice a week, and the best time for him is between 7 a.m. and 8 a.m. Bearing in mind nothing else but your own "feeling best" rhythm, how do you think you would perform?

   Would be in good form **(4)**   Would find it difficult **(2)**

   Would be in reasonable form **(3)**   Would find it very difficult **(1)**

10. At what time in the evening do you feel tired and, as a result, in need of sleep?

    8 p.m. to 9 p.m. **(5)**   12:45 a.m. to 2 a.m. **(2)**

    9 p.m. to 10:15 p.m. **(4)**   2 a.m. to 3 a.m. **(1)**

    10:15 p.m. to 12:45 a.m. **(3)**

11. You wish to be at your peak performance for a test that you know is going to be mentally exhausting and will last for two hours. You are entirely free to plan your day. Considering only your own "feeling best" rhythm, which test time would you choose?

    8 a.m. to 10 a.m. **(6)**   3 p.m. to 5 p.m. **(2)**

    11 a.m. to 1 p.m. **(4)**   7 p.m. to 9 p.m. **(0)**

12. If you went to bed at 11 p.m., at what level of tiredness would you be?

    Not at all tired **(0)**   Fairly tired **(3)**

    A little tired **(2)**   Very tired **(5)**

13. For some reason you have gone to bed several hours later than usual, but there is no need to get up at any particular time the next morning. Which one of the following events are you most likely to experience?

    Will wake up at the usual time and will not fall back asleep **(4)**

    Will wake up at the usual time and will doze thereafter **(3)**

    Will wake up at the usual time but will fall back asleep (2)

    Will not wake up until later than usual **(1)**

14. One night you have to remain awake between 4 a.m. and 6 a.m. in order to carry out a night watch. You have no commitments the next day. Which one of the following alternatives will suit you best?

    Would not go to bed until the watch was over **(1)**

    Would take a nap before and sleep after **(2)**

    Would take a good sleep before and nap after **(3)**

    Would take all sleep before the watch **(4)**

15. You have to do two hours of hard physical work. You are entirely free to plan your day. Considering only your own "feeling best" rhythm, which of the following times would you choose to do the work?

    8 a.m. to 10 a.m. **(4)**   3 p.m. to 5 p.m. **(2)**

    11 a.m. to 1 p.m. **(3)**   7 p.m. to 9 p.m. **(1)**

16. You have decided to engage in hard physical exercise. A friend suggests that you do this for one hour twice a week, and the best time for him is between 10 p.m. and 11 p.m. Bearing in mind nothing else but your own "feeling best" rhythm, how well do you think you would perform?

    Would be in good form **(1)**   Would find it difficult **(3)**

    Would be in reasonable   Would find it very
    form **(2)**   difficult **(4)**

17. Suppose that you can choose your own work hours. Assume that you worked a five-hour day (including breaks) and that your job was interesting and paid by results. What time would you finish?

    5 a.m. to 8 a.m. **(5)**   3 p.m. to 5 p.m. **(2)**

    9 a.m. **(4)**   6 p.m. to 4 a.m. **(1)**

    10 a.m. to 2 p.m. **(3)**

18. At what time of the day do you think you reach your "feeling best" peak?

    5 a.m. to 7 a.m. **(5)**   5 p.m. to 9 p.m. **(2)**

    8 a.m. to 9 a.m. **(4)**   10 p.m. to 4 a.m. **(1)**

    10 a.m. to 4 p.m. **(3)**

19. One hears about "morning" and "evening" types of people. Which one of these types do you consider yourself to be?

Definitely a morning type **(6)**

Rather more a morning than an evening type **(4)**

Rather more an evening than a morning type **(2)**

Definitely an evening type **(1)**

Results

To determine your score, find the point value next to each of your answers. Add them together to get your total number of points to determine your bird type. Here's what your score means:

Score	Result
16–30	Super Owl
31–41	Owl
42–58	Hummingbird
59–69	Lark
70–86	Super Lark

Super Owls and Owls

You, like about 20 per cent of the population, have a hard time waking up early and are most energetic in the evening. The degree to which your behaviour manifests itself in your propensity to stay up late determines whether you are an extreme Super Owl or a more moderate Owl.

But no matter where you fall on the spectrum, all owls tend to be intuitive, emotional and creative. You are most likely to work in the arts, medicine or technology. You crave novelty and can often be a risk taker. Socially, you can be an introvert – often showing up late to the party and preferring to observe. But once someone gets to know you, they'll find you to be a deeply loyal friend.

Some famous Super Owls and Owls include former President Barack Obama, writer Fran Lebowitz, Microsoft cofounder Bill Gates and musician Elvis Presley.

Hummingbirds

You are among the most popular group of sleep birds: 70 per cent of people are Hummingbirds, meaning you are more active during the day and restful at night. While this may be the most common chronotype, it does not mean all Hummingbirds are alike. Depending on where you score, your behaviour may be similar at times to an Owl or it could be more like that of a Lark.

Whatever your extreme, you most likely need 8 hours of sleep, and feel the need to nap in the late afternoon, as sunlight has a big influence on your circadian rhythm. But while you take comfort in the familiar, you have the ability to occasionally change your schedule – so you can stay out past your bedtime for a party or wake up to watch the sunrise. Hummingbirds tend to be extroverted and open-minded. You can also be cautious and tend to avoid conflict. Friends, and you have many, would say that you are quick to share a funny story, easygoing, fun and happy.

Super Larks and Larks

You are the rarer bird that wakes up early and energized, but by evening has totally run out of steam. About 10 per cent of people are Larks. And again, the degree to which your behaviour manifests determines if you're the more extreme Super Lark, with a propensity to need sleep much earlier in the evening or wake up earlier in the morning, than a more moderate Lark.

But for all Larks, a bright-eyed, go-get-'em eagerness might be their defining characteristic. You are often successful in business and school due to your analytical mind and aversion to risks. You enjoy being a leader and may be a CEO or entrepreneur. In business or in your personal life, if you make a plan, you stick to it – and this is especially true of your health and fitness routine. Socially, you may lag behind a bit, as you tend to go to bed early, but you're also up early to catch up on what you missed.

Some famous Super Larks and Larks include editor in chief of American *Vogue* Anna Wintour, former First Lady Michelle Obama and former New York City mayor Michael Bloomberg.

Whatever kind of sleep bird you identify with today, know that it can change. This can be attributed to changes in your circadian rhythm, which, again, along with sleep monitors body temperature and hormonal changes. And it is particularly important for women, as both our body temperature and our hormones fluctuate from menstruation to pregnancy to menopause – disrupting our rhythm and changing our sleep needs. And here's a fun fact: we all grow more Lark-like with age.

Knowing your chronotype allows you to not only be more productive, but also strategize to sleep better.

If you don't have trouble falling asleep and usually wake up rested, you are most likely in sync with your circadian rhythm and optimizing your autophagy. But many of us suffer social jet lag. This happens when our chronobiology doesn't mesh with our responsibilities. For example, a Lark may feel tired at 5 p.m., but still need to work until 7 p.m.

And no matter your chronotype, you can make simple shifts to optimize your circadian rhythm, sleep better and activate your autophagy.

Customi-Zzz Your Sleep

Birds of a feather don't have to flock together! The Glow15 plan helps you tailor your sleep for your bird type. It is, in a sense, maximizing your circadian rhythm to get it back in tune and synchronizing it with your autophagy so you have more energy, perform better, feel better and look better, too. Here's how to improve your sleep and then make adjustments depending on your chronotype.

Let in Light

Sunlight helps us to adjust our body clocks every day. It works like this: if you get lots of light in the morning, your body clock speeds up, so you will want to go to sleep earlier and wake up earlier. But if, instead, you get lots of light in the late afternoon, the opposite happens – your body clock slows, so you will want to go to bed later and wake up later.

Here's how each bird type should let in light:

Super Owls and Owls: You should get lots of light in the morning. Exposure to daylight can keep you more alert, helping you to jump-start your day. The best way to do this is to go out into sunlight early in the day. Another option is to use a light box with about 10,000 lux of light, which is the equivalent of early daylight. Also, keeping your shades open at night will allow natural light to wake you in the morning.

Hummingbirds: Keep on doing what you're doing, because all is well, and you don't need to change a thing.

Super Larks and Larks: Getting lots of light in the afternoon and around sunset can slow down your body clock and help you stay up later into the night. Ideally, it best to get natural sunlight. But if that's not possible, try a light box with approximately 10,000 lux – the equivalent of a clear morning light. Try keeping your shades closed at night – "blackout" shades can be a great investment to help you sleep.

Set Your Bedtime

You should be going to bed and waking up at the same time every day – yes, that includes weekends. Your internal clock doesn't change the way it keeps time just because it's Saturday! Of course, there will always be exceptions, but in general, you want to always be in sync – which means orchestrating your sleep cycles and your circadian rhythm for the most efficient, quality sleep.

Your sleep cycle is about 90 minutes long – and cycles through five stages of sleep: Stage 1 is the transition from wakefulness to light sleep. Stage 2 is slightly deeper, but you can still awaken easily. Stages 3 and 4 are when sleep becomes much deeper. And the final stage, REM (rapid eye movement), is where the deepest sleep happens, and restoration – from your nervous system to how you process information to how you store memories – takes place.

For sleep to be restorative, you need several complete sleep cycles every night; most people require four or five complete cycles. Some women feel powerful and productive after 6 hours of sleep, or four 90-minute sleep cycles; others claim to be at their best with more or less sleep. This depends upon the length of your personal sleep cycles. Generally, Owls and Super Owls feel better with four sleep cycles, while Hummingbirds, Larks and Super Larks feel better with five.

To find your best bedtime, multiply the time of one cycle (90 minutes) by your desired number of cycles. Let's say you are a Hummingbird – you want five cycles. So multiply 90 by 5 and you get 450 minutes, or 7.5 hours.

But, again, we want the best-quality sleep – the kind of sleep that leaves you feeling refreshed and looking rejuvenated in the morning. And to get that, you also have to factor the time it takes you to fall asleep. For most of us, that is between 20 and 40 minutes. Add that to the total number of your cycles. Owls and Super Owls usually take longer to fall asleep, around 40 minutes, while Hummingbirds, Larks and Super Larks are more likely to fall asleep in 20 minutes. So, for example, if you are a Hummingbird, you would add 20 minutes to your 450 minutes for a total of 470 minutes.

Since most of us have a wake-up time that's determined by kids or work (this is your socially determined wake-up time), you can count backward from the time you need to wake up. For a hummingbird to wake up at 7 a.m., you subtract 470 minutes from that time, for a bedtime of 11:10 p.m.

The goal is to wake up naturally a few minutes before your alarm. When you can do that, you've found your bedtime. If you wake up way before your alarm or have a hard time getting up when your alarm goes off, don't despair – remember, each of our internal clocks keeps time slightly differently. If you wake up too early, try shifting your bedtime forward, and if you're waking up late, shift it backward. You may have to experiment a bit. But you *will* find the sweet spot where you wake naturally according to your own rhythm.

Here's a guideline for when each type of bird should set their bedtime:

Super Owl: 1 a.m. Super Owls will take about 40 minutes to fall asleep. But they usually need only four sleep cycles. So, to wake up at 8 a.m. – which is likely the latest you can wake without suffering the aforementioned social jet lag – you need to subtract the 400 minutes, or four 90-minute cycles plus 40 minutes to fall asleep, which gives you a bedtime of 1 a.m.

8 a.m. - 400 minutes (40 minutes to fall asleep + 4 cycles) = 1 a.m. bedtime

Owl: Midnight. Owls will also take about 40 minutes to fall asleep and need only four sleep cycles. But an Owl can wake up a little earlier than a Super Owl. So, to wake up at 7 a.m., Owls should subtract the 400 minutes – or four 90-minute cycles plus 40 minutes to fall asleep – which gives you a bedtime of 12 a.m.

7 a.m. - 400 minutes (40 minutes to fall asleep + 4 cycles) = 12 a.m. bedtime

Hummingbird: 11:10 p.m. Hummingbirds take about 20 minutes to fall asleep and need five sleep cycles. So, to wake up at 7 a.m., Hummingbirds should subtract the 470 minutes – or five 90-minute cycles plus 20 minutes to fall asleep – which gives you a bedtime of 11:10 p.m.

7 a.m. - 470 minutes (20 minutes to fall asleep + 5 cycles) = 11:10 p.m. bedtime

Lark: 10:10 p.m. Larks take about 20 minutes to fall asleep and need five sleep cycles. So, since a lark naturally wakes early, 6 a.m., she should subtract 470 minutes – or five 90-minute cycles plus 20 minutes to fall asleep – which gives you a bedtime of 10:10 p.m.

6 a.m. - 470 minutes (20 minutes to fall asleep + 5 cycles) = 10:10 p.m. bedtime

Super Lark: 9:10 p.m. Super Larks also take about 20 minutes to fall asleep and need five sleep cycles. But you naturally wake by 5 a.m., so subtract 470 minutes – or five 90-minute cycles plus 20 minutes to fall asleep – which gives you a bedtime of 9:10 p.m.

5 a.m. - 470 minutes (20 minutes to fall asleep + 5 cycles) = 9:10 p.m. bedtime

How to Fly with a Different Flock

- If you are an **OWL** who needs to operate on a **HUMMINGBIRD** schedule, then you would use light in the a.m., right after you wake up, to shift the circadian cycle earlier.

- If you are an **OWL** and you need to be a **LARK,** it may make more sense to get less sleep (three cycles) and take a 90-minute nap (one full cycle). You will need light in the a.m. for sure.

- If you are a **LARK** who wants to be a **HUMMINGBIRD,** you could use light therapy in the early evening. Your body will wake you up naturally and slowly shift over time.

Now, if your life dictates that you need to be up much earlier or later than what is preferred for your bird type, you can try to hack your chronotype. It *will* take a bit of time, as the best way to do this is gradually, in 15-minute increments in either direction. So, Owls and Super Owls who want to be more Lark-like should set their bedtime 15 minutes earlier per night. Larks and Super Larks who want to be more Owl-like should set their bedtime 15 minutes later per night. Stick with just the 15-minute change for at least three nights before further adjusting your bedtime. Once you have got used to that change – and you feel your circadian rhythm is in tune with the change – shave off another 15 minutes. You can continue to do this until you reach your desired sleep schedule. Also, note that it's normal to wake up a bit dazed or befuddled. That sleepy feeling should go away within about 25 minutes of waking. (Hummingbirds do not need to make any adjustments.)

What All Birds Should Do Before Nesting

You know how most people think of sleep? That it's just something that should naturally happen. We should be able to shut our eyes and lull into a state of blissful rest. But with a spinning brain and a stressful life, that's not the way it works. We can't just hop into bed, shut off the lights, pull up the covers and expect that we'll be dreaming of desert islands in no time.

Smart sleep requires smart thinking – no matter what type of bird you are. Here is how you can get in sync with your circadian rhythm and optimize your autophagy:

Power Down. I know how tempting it is. We snuggle up in bed, grab our phones and scroll through our feeds to see the latest bits of craziness that our friends are up to. But exposure to the blue and white light given off by virtually all electronic devices will make it more difficult to sleep. So for optimal sleep prep, that means no technology for 90 minutes before bed. If you can't live without a pre-bed Netflix binge, don't worry; there is software that will reduce the blue hues from your screen. Orange-tinted glasses or orange lightbulbs will also do the trick.

Dr Breus suggests the Power Down Hour: set an alarm for the hour *before* you go to bed, to remind you to start the process of "powering down". In the first 20 minutes, take care of all the things you simply must do before bed: make lunches for the next day, lay out clothes, put away dishes (in my house it's getting backpacks ready and finding shoes) – the things that will nag at you if you don't do them. In the second 20 minutes, take time for personal hygiene: brushing your teeth, washing your face, changing into your nightclothes. A hot shower or nice bubble bath can be great just before bed. The change in your core body temperature when you get out of the shower or tub will help your body start to relax. The final 20 minutes are a time for relaxation: lower the lights, do some simple stretches, read or listen to some relaxing music.

Cool down. Ideal room temperature is about 18°C. The cooler your body, the better you will sleep. That decrease in core body temperature helps align your body with circadian rhythms, so the effect is better and more comfortable sleep.

Null the noise. Silence is sleepy, but many of us live in large, noisy cities, or have family members unwittingly making a racket just when we're about to drop off. Thankfully, the "colours" of noise can help – yes, colours. In fact, I recently learned that many noises can be characterized by a sonic hue determined by two things: the collection of frequencies or pitches in the sound (also called a spectrum) and the amplitude or intensity of the frequencies. The colours are supposed to correlate with the general characteristics of colours of light. You've probably heard of white noise, which includes a random set of frequencies at the same intensity. This combination often sounds like some form of static. Pink noise has higher intensity at lower pitches, sounding more like rain falling. And brown noise goes a little further, dampening the highest frequencies and producing a sound like waves crashing. All of these noises can help you sleep – many people find the deeper, richer brown noise helps lull them best, while others find pink noise more relaxing. While you can buy a noise machine, there are also many apps you can download for free on your smartphone or computer to help you determine which noise colour is best for you.

Take a beauty bath. To boost your beauty sleep and outsmart ageing, try adding a hot Epsom salts bath to your sleep schedule. Two hours before bed, soak in the tub for 20 to 30 minutes. Epsom salts break down into magnesium and sulphate in hot water. The minerals are absorbed through your skin and help aid in detoxification. The heat also helps your circadian rhythm. This is because your natural drop in body temperature is balanced with the heat of the bath. Then, when you get out of the hot bath, your temperature will plummet again, making you all the more ready for sleep. You can also try a shower, although it may not be quite as helpful. Plus, there's some new research that suggests that bathing for an hour can have some similar positive effects as exercise. According to the study author, Jennifer Wider, MD, this happens through "passive heating". When you're heated, like in saunas, baths or exercise, heat shock proteins can become elevated, and that helps control blood sugar and insulin function.

What If You Still Can't Sleep?

Sleep, as you know, is a complex issue. And even when you do everything you can to make the environment right for sleeping, you can still have problems – and that's a problem, because too little sleep can cause all kinds of health issues (such as heart disease, depression and strokes) that will accelerate ageing. Many women, as you likely know, say that psychological distress is the source of sleep issues. One of the things that you can do is start a "worry journal", where you write down your problems and potential solutions for them. You pack it up at night, clearing your mind for sleep, knowing that you can address the issue the next day with a clear plan in place. Some more specific tips for common sleep problems:

You can't fall asleep. Try bright light for two hours in the morning, which can help adjust your circadian rhythm so you fall asleep faster.

You can't stay asleep. Use bright light therapy between 7 and 9 p.m. to delay the timing of your circadian rhythms.

Jet lag. Adjusting your watch to the new time zone when you board the plane has been shown to help ward off jet lag. Exposing yourself to light can help, too. On an eastward flight, when arriving in the early morning,

What to Do During the Day to Improve Your Nights

Having energy throughout the day isn't just about how well you sleep; it also means that you do other things, as well. Eating well, being active and keeping stress down also influence how you live and how you rest. In my life, I've found that two other easy practices have made me sleep better at night – and clear my mind and improve my performance during the day. I start the day with a short meditation and, when I can, I squeeze in a short nap.

Meditate

Meditation has been shown to improve sleep, as well as to lower blood pressure and stress. Try doing it in the early morning, when your blood pressure and stress hormones are at the highest. I do it as soon as I wake up to clear my mind and get me started on my day.

To meditate, just sit in a quiet place, breathe in a relaxed way and try to focus on one word or phrase over and over. This is called a "calming focus". You can use a breath, a sound ("om"), a short prayer, a positive word (such as "relax" or "peace") or a phrase ("breathing in calm, breathing out tension"; "I am

relaxed"). If you choose a sound, repeat it aloud or silently as you inhale or exhale. If you find your mind wandering, take a deep breath or say to yourself "thinking, thinking" and gently return your attention to your chosen focus. Start by doing it for just a few minutes; you can build up to 5 to 10 minutes, if you like.

Nap

In some circles, *nap* is as evil a word as *doughnut*. That's nonsense! As I mentioned earlier, studies have found that our circadian rhythms dip during the day, showing that we're actually meant to have two sleep periods: the obvious long one at night and then another in the early afternoon, which is when our energy naturally dips lower than usual and we have a harder time focusing. This is why an afternoon nap can significantly increase mental alertness and improve mood. The ideal nap map:

- **Keep it to 10 to 20 minutes,** the ideal time that will help you awaken easily and feel refreshed and recharged without feeling groggy.
- **Be in a seated position but slightly reclined.** This will

121

prevent you from falling into a deeper sleep, which can leave you feeling lethargic when you wake up.

- **Use an alarm.** Set an alarm for 10 to 20 minutes. In addition, you can hold a pen or pencil in your hand. It will drop about 10 minutes after you fall asleep and should wake you, too.
- **Find a quiet, dark place.** Ideally, you want a private place for your nap. If possible, turn off the lights or draw the curtains to make it dark.
- **Drink caffeine.** Contrary to what you might think, drinking a small cup of AutophaTea (page 199) or coffee right before you nod off will not keep you up! In fact, it will give you more energy when you wake up, as it may help lessen the effects of sleep inertia.

you should avoid bright light – wear sunglasses until noon. After noon, take your sunglasses off and get as much direct sunlight as possible, especially between 1:30 and 4:30 p.m. If you're stuck inside all afternoon, take hourly sunshine breaks of 10 to 15 minutes each. On a westward flight, wear sunglasses and try to block out light during the flight. Keep your shades on until the last two hours of the flight. Then take them off and get as much light as possible via light from the window or close-up screen exposure. Get as much direct sunlight as possible as soon as you arrive.

Shift work. When you often change the times when you work, especially during night hours, it disrupts your circadian rhythms. For night workers, you need to delay your circadian rhythm. Get exposed to bright light during a night shift for three to six hours. You can also try doing it for 20 minutes every hour during the night shift.

One of the best benefits of a good night's sleep is waking up feeling revived and looking it, too. There is a reason it's called beauty sleep! As your autophagy is activated, it helps to repair and renew your cells, allowing your skin to refresh overnight. In the next chapter, find out the Glow15 way for your skin to look great all day. It's everything you need to know about how to glow.

GLOW15 SUCCESS STORY

"I'm sleeping better now than I did before children"

TAMMY

"Glow15 is phenomenal! I'm a mum of two– my daughter is four and my son is six. Both were C-sections, and I've had a very hard time losing the baby weight. Before I got pregnant, I was petite and fit, and I had no idea it would be so difficult to get my body back. I've tried so many diets, but nothing worked. Then **I started Glow15 – and now my clothes are big on me! I lost over 7 pounds in the first 15 days, and they keep coming off.** My midsection looks much better and I am starting to see my old waist. During the first few days I was a little tired, but now **I have more energy than I've had in years.** I feel great. I feel like a younger version of myself. And I think the 'Glow to Sleep' plan is the real reason. I discovered that I'm an 'owl'. I always knew I was a night person, but I had no idea that meant I should do anything special – like getting more light in the morning or changing my bedtime. Now, I can't believe **I'm actually sleeping better now than I did before children!** It worked so well, **I wake up rested and refreshed.** Maybe it's this newfound energy that helped me finally lose the weight. I think when I was struggling to get it off I was a little bit down, but now because of Glow15 I feel much more positive and happy."

The Three B's of Better Sleep

Sleep is not an on/off switch, but more like when you slowly take your foot off the accelerator pedal and slowly push down on the brake. It's a process that takes a little time and requires various things to happen to fall asleep. If you fall asleep as soon as your head hits the pillow, it probably means that you're sleep deprived. To nudge your sleep in the right direction, consider the three B's: Bedroom, Body and Brain.

1. Bedroom

It should exude tranquility, not clutter and stress. You can turn it that way by doing these things:

Change the bulbs in your bedside table lamp to 40 watts, put a nightlight in your bathroom and in the hallway leading to it, and consider an eye mask. Why? Light tells your brain it is morning and to stop producing melatonin.

Use earplugs or a sound machine to block out unwanted noises.

Assess your mattress to make sure it's comfortable. If you have neck or back pain, or your mattress is more than ten years old or your pillow more than three years old, you may need to replace them.

Clean up and cool it down. Eliminate clutter, and lower your core body temperature. The heat from a warm bath or shower will raise your core body temp, and then when you get out, it falls. This is a signal to your brain to begin producing melatonin.

2. Body

Are you doing what you need to do on a physical level to get to sleep more quickly? Regular exercise has been shown to promote deep sleep and shorten the time it takes to fall asleep. Research shows that people in pain have a hard time sleeping as well. See your doctor and get the pain under control to get a quicker and better night's rest.

3. Brain

Continuing to think about stressful events, times or thoughts will cause arousal, which will make it harder to sleep. You can try relaxation music or apps. Dr Breus also suggests counting backward from 300 by 3. It is no easy task, and it is so boring that it should put you to sleep.

Chapter 8

Glow Younger

Nourish Your Skin Cells and Outsmart Wrinkles

Is there any better compliment than "You're glowing"? It's the ultimate praise because it implies that you look youthful, fresh and filled with health. But when you're tired or ill, you may hear that you look "pale" or "worn out".

The quest for youthful, glowing skin is not superficial. Your appearance is tied to so many other factors in life. It plays a role in first impressions, which make all the difference in all types of encounters (remember my story of Clive and how he rejected me once he got a glimpse of my skin covered in eczema), from business to personal. The way you look plays a role in factors that influence how you live day-to-day: confidently, sensually, powerfully and happily.

More than that, skin isn't just about how you look or how appealing you are to others. Your skin – your largest organ – is a direct reflection of your health and well-being. Dry, sun-damaged, dull skin and fine lines and wrinkles can be visible signs of your body's degradation. But when your skin glows it radiates inner beauty and better health.

Beauty, as the saying goes, is only skin-deep. But the importance of skin goes a lot deeper. In this chapter, I'll show you how to activate your youth to get healthy, glowing skin by activating your autophagy. Find out how to do this through nutritional ingredients and cutting-edge technology.

Why Your Glow Goes Away

The biggest threat to your health, and the health of your skin, is ageing. The Inevitable and Accelerated Agers (see page 32) leave an invisible impact on your body and a visible impact on your skin in the form of dryness, loss of collagen and elasticity, fine lines and wrinkles, and age spots. Dry skin (or dry patches of skin) is often one of the first signs of ageing, but it also exacerbates most of the other ones. This is why dermatologists recommend we start using a moisturizer in our early twenties.

But what dermatologists didn't realize until recently is the relationship between the Accelerated Agers, autophagy and the skin. A breakthrough study presented at the American Academy of Dermatology conference showed that Accelerated Agers such as UV irradiation and free radical inducers can suppress autophagy. This is incredibly significant because we now understand that not only does autophagy naturally decline with age, but Accelerated Agers further suppress autophagy in the skin, and therefore further hinder your cells' ability to clean up and repair the damage that causes visible signs of ageing. As a result, we see an accelerated decrease in vital collagen production and a delay in cell turnover, which contributes to dry and dull skin, fine lines, wrinkles, loss of firmness and thinning of the skin, all commonly referred to as signs of ageing.

In addition to the known Accelerated Agers you've already read about in this book, there is another category of Accelerated Agers that may be the reason your glow is going away, and these are likely hiding in your bathroom cabinet. Many products you use daily, even those that claim to be anti-ageing, can make your skin look older and age faster than it should. This is because the ingredients can contribute to the waste and damage in your cells and slow autophagy. Common ingredients that can damage cellular components and hinder autophagy are in the following table.

Formaldehyde

Often used in cosmetics, lotions, creams, nail polish, eyelash glue, hair gel and deodorants.

An article published in *Contact Dermatitis* found that nearly 20 per cent of products contain formaldehyde or formaldehyde-releasing preservatives. This chemical is linked to allergies, chest pain, chronic fatigue, depression, dizziness, ear infections, headaches, joint pain and loss of sleep, and can trigger asthma.

Note that on a label, formaldehyde may be listed as *formalin,* or you may have to play detective and scrutinize the label for one of the following: *DMDM hydantoin,* found in more than 2,000 products; *diazolidinyl urea,* found in about 2,000 products; *imidazolidinyl urea,* found in more than 700 products; *quaternium-15,* found in about 400 products; and *hydroxymethylglycinate,* found in more than 300 products.

Mineral Oil and Petrolatum

Often found in lip balm, lipstick, day and night creams, lotions and ointments

These petroleum derivatives coat the skin like clingfilm, limiting the skin's ability to breathe. Mineral oil doesn't absorb into the skin – its molecular size is simply too big. As a result, it stays on the surface. Instead, look for natural emollients. Wax esters such as jojoba, candelilla, and carnauba are great oil-soluble options, as are cocoa butter and shea butter. I really like beeswax because it helps to prevent water loss and has the same moisturizing properties as mineral oil, but doesn't pollute and clog pores and cause breakouts.

Oxybenzone

Used in sunscreens, lip balm and moisturizers to protect against UV rays

This chemical has been linked to hormone disruption, cancer and low–birth weight babies. Oxybenzone can cause an allergic reaction in many people when it is exposed to the sun. Quick tip: to avoid oxybenzone but still get the benefit of sunscreen, look for products that contain zinc oxide or titanium dioxide.

Parabens and Phthalates

Found in cosmetics, creams, hairspray, nail polish and perfumes, these preservatives keep out bacteria and prolong the life of products

Parabens and phthalates are preservatives used to keep out bacteria and prolong the life of products, but they are thought to be potentially carcinogenic. They were banned by the European Union in 2003. Parabens can be listed under chemical names like *methylparaben, propylparaben, isoparaben* or *butylparaben*. If you see the words "fragrance" or "parfum" on a label, that almost always means phthalates. To ensure your products are free of phthalates, look for the following claims: "no synthetic fragrance", "scented with only essential oils" or "phthalate-free."

Sodium Lauryl Sulphate (SLS)

A common foaming agent found in personal care products, including shampoos, cleansers and toothpastes

SLS is a harsh yet effective cleansing agent that strips away oil and other gunk. But this powerful stripping quality also makes SLS one of the primary skin irritants used in body care products today. Its caustic lathering agent can destroy the lipid layers that keep the skin smooth and supple. The *Cosmetic Ingredient Review* recommends you chose less than 1 per cent SLS to avoid irritation. An even safer option is to choose "SLS-free" products.

So not only do Accelerated Agers suppress autophagy, toxins hiding in your bathroom cabinet can cause a buildup of toxic waste in your cells, further hindering autophagy. That, coupled with the fact that your autophagy naturally declines, makes it almost impossible for your cellular cleansing cleanup crew to do its job efficiently. The buildup of junk and damage to your cells leads to signs of ageing in your skin like dryness, dullness, wrinkles and discolouration. In other words, your glow begins to go away, and for good reason, but you can get glowing again. Amazingly a 2014 study carried out by CACI Microlift found British women start worrying about the ageing process at the age of 24, with three years of anti-wrinkle cream purchases already under their belts.

How to Get Glowing

Even though you're getting older, you can still look younger.

Dr Nicholas Perricone is a board-certified clinical and research dermatologist, a world-renowned leader in anti-ageing skincare and author of several *New York Times* best-selling books based on his passion and knowledge of the effect of antioxidants on our skin's health. A line at the end of his book *The Wrinkle Cure* has stood out to me since the first time I read it many years ago, and it's one of the driving factors that encouraged me to pursue autophagy activation and its impact on the skin. He says, "I'm convinced that the antioxidant revolution is just the beginning. Each triumph brings us one step closer to making the fantasy of eternal youth and beauty a realistic and obtainable goal."

For so long, I was consumed with the idea that antioxidants were the be-all and end-all of anti-ageing. But this line from Dr Perricone kept me wondering, "What's next?" After years of research and discovery, I'm now convinced it is autophagy. Autophagy begins where antioxidants leave off. While antioxidants help to prevent damage due to the Accelerated and Inevitable agers, autophagy does what antioxidants can't — it acts to remove and repair the skin's already damaged cellular components.

What's more, you have the power to boost your autophagy, rid your cells of damage and toxins, and repair your skin, so that it looks and behaves younger. New and very exciting research suggests that the stimulation of autophagy can positively influence each layer of skin to repair the visible and invisible signs of ageing. I worked extensively with my good friend, celebrity dermatologist Dr Dendy Engelman, to better understand the role of each layer of skin, how it's impacted by Accelerated and Inevitable agers and how each layer may be affected by autophagy activation. Let's dive in.

Epidermis: This is the topmost layer of skin that protects your body from the elements. It guards against Accelerated Agers like the sun, pollutants and chemicals and acts as a biological shield against other invaders like insect bites and stings. And this is the layer where lines, age spots and discolouration form. Skin cell turnover happens in the epidermis; this is the natural exfoliation process in which dead cells should slough off naturally every twenty-eight days or once a month. But as you age from both the Accelerated and Inevitable agers, this sloughing process dramatically slows down, as slow as once every forty to fifty days, which makes your skin dry, dull and older looking than it should. In addition, the epidermis holds in the skin's water and precious oil. Fine lines are formed when the epidermis loses its ability to hold onto this moisture, causing your skin to thicken and become rough and dry. In addition, this outer layer is where skin-colour-producing melanin is formed in cells. Age spots and uneven blotches occur when the epidermis overproduces melanin.

A conference of the American Academy of Dermatology revealed that autophagy activation can lead to increased cell turnover, helping to eliminate the dryness that causes fine lines. Increasing this cell turnover revitalizes the overall health and youthful appearance of your skin. Boosting autophagy can also stop the overproduction of melanin that causes age spots, as indicated in the journal *Experimental Dermatology*. As mentioned above, age spots form when melanin, which is what produces skin pigment, is overactive. UV light accelerates the production of melanin and thus the formation of dark spots. Slowing down melanin production helps slow discolouration.

Dermis: This is the middle layer, consisting of connective tissue, capillaries, nerves, hair follicles, and oil and sweat glands. Collagen and elastin fibres weave through the dermis, giving the skin firmness, elasticity, and structure. By the time you reach age 25, you produce about 1 per cent less collagen per year. That loss gets even greater as you age. When you reach your forties, you may have 10 to 20 per cent less collagen than you did in your twenties. And at fifty, you may have up to 50 per cent less collagen than you did in your twenties. Collagen depletion leads to the formation of wrinkles, sagging skin, easy bruising and loss of elasticity. Also noteworthy: wounds take longer to heal with a lack of collagen.

Stimulating autophagy in this middle layer can lead to an increase in collagen production, according to the journal *FASEB*. Activating autophagy also slows the process of cellular senescence. Senescence is the cell condition of deterioration with age – the loss of cells' power to divide and replicate. Slowing senescence makes it possible to preserve the function of skin tissues and ward off the visible signs of ageing.

Hypodermis or Subcutaneous Layer: This is the bottommost layer, often referred to as the "fat" layer because it is composed mostly of fat and connective tissue lying atop your muscles. This is the layer that helps regulate body temperature and protects your muscles and bones. It also keeps your skin smooth and plump. As we age, the fat pads that once gave us fullness through the mid-cheeks – or the "apples" – naturally start to disappear and the connective tissue loses its resilience, leading to sagging skin.

Autophagy plays an essential role in the homeostasis of adipocytes (fat cells). There is also evidence that autophagy plays a role in the interconversion of brown and white fat cells. In general, younger people have more brown fat cells. Brown fat cells convert excess sugars and fatty acids into heat for temperature regulation. In contrast, white fat cells convert excess sugars and fatty acids into storage fatty acids. With increased age and deteriorating autophagy, the balance between white and brown fat cell formation may be compromised, which may have broader impacts on health.

A Remarkable Discovery

While science shows that stimulating autophagy can improve cellular function, the application of autophagy in skincare is only just being realized. One of the most respected and credentialed dermatologists studying autophagy and the skin is Richard Wang, MD, PhD. I've been fortunate enough to work with him on the development of Glow15.

Dr Wang regularly sees patients and is an assistant professor of dermatology at one of the most prominent centres for research in the field of autophagy. He is among a select few scientists who have conducted significant research related to autophagy and skin.

The 2016 Nobel Prize-winning research around autophagy inspired Dr Wang and me to revolutionize the approach to anti-ageing nutritional skincare ingredients. While traditional ingredients largely focused on the *external* signs of ageing, our approach was to combat what we now know as the fundamental or root cause of ageing – *the decline of autophagy*. Dr Wang and I hypothesized that if we could boost autophagy levels to what they were in our youth, our skin will not only repair damage with the same efficiency as young skin, but cellular health would also be prolonged and signs of ageing could be delayed and even reversed.

With my commitment to pure and efficacious nutritional ingredients and Dr Wang's research and understanding of the optimal conditions for autophagy activation, we went on a quest to identify ingredients that not only could activate autophagy, but were also proven to protect, repair and nourish the skin.

The scientific literature is littered with claims about different ingredients that can activate autophagy. After carefully researching hundreds of ingredients over the course of two years, I brought the leading ingredients to Dr Wang, and through his research we landed on powerful nutritional ingredients that together activated autophagy over multiple cellular pathways when tested in-vitro. This is remarkable and significant because the complex process of autophagy occurs across multiple cellular pathways, and our nutritional ingredients activate each of them. The key ingredients are ceramides, trehalose, polyphenols and caffeine.

Putting the Nutritional Ingredients to the Test

Dr Wang's clinical results, coupled with the transformative changes to my skin (you know I'm always my own human guinea pig), were so inspiring that I decided to share the nutritional complex with all of the women in the Jacksonville University study.

After sixty days of applying the autophagy-activating nutritional ingredients every morning and night, 100 per cent of the participating women reported a dramatic improvement in their skin's youthfulness, citing a visible reduction in fine lines and wrinkles as well as a reduction in pore size. Additionally, 85 per cent found their skin to be firmer and plumper, and their complexions were more youthful and smoother. And 88 per cent of the participants found their complexions were visibly more even.

Facial photographs, assessed by an independent board-certified dermatologist, confirmed the subjective self-report findings of improvements in the women's skin health. In addition, digital photographic analysis using the VISIA facial imaging system showed statistically significant improvements, including improvements in skin texture and reductions in ageing markers like brown spots and facial wrinkles.

The most rewarding part of the process was hearing from the participants themselves. They repeatedly told me how much happier and better they felt about their appearance. Many of them described their skin as "radiant," "brighter" and "glowing."

I believe in a multifaceted approach to fight skin ageing – treating both the visible and invisible signs. So, in addition to diet, exercise and sleep modifications, working with a team of research scientists, cell biologists and dermatologists we have taken the autophagy-activating nutritional ingredients that Dr Wang and I identified and transformed them into a first of its kind range of nutritional skincare. We can't wait to share the range with you, but in the meantime we want to share with you some home remedies you can make yourself in your kitchen.

Targeted Topicals to Glow

Some of the best nutritional ingredients to activate autophagy are part of Glow15. While ingesting them helps in the systemic stimulation of autophagy, applying them topically can repair and rebuild the very scaffolding on which your skin sits, so the results are long-lasting and continue to get better and better over time.

The nutritional ingredients that maximized autophagy activation during our clinical research and produced astonishing results for the women in the Jacksonville University study are listed below. These ingredients will activate your skin's natural youth-promoting cellular activity, keeping you healthy and beautiful – both inside and out.

Ceramides

These sphingolipids (see page 70) lock in moisture to help plump the skin. Found in the epidermis, they are a crucial component of your skin's barrier against pollutants and harmful bacteria. Like most things, your skin's natural ceramide production decreases as you age, which can lead to dryness, fine lines and wrinkles. Decreased ceramide levels are also associated with skin conditions such as atopic dermatitis, eczema and psoriasis.

The good news: topical ceramides can replenish your natural lipid loss, help restore youthful moisture, protect skin from irritants and fortify your skin's natural barrier. In a Japanese study, eyelids, a key indicator of facial ageing, treated with a ceramide gel for four weeks showed a significant increase in water content. Numerous clinical studies have also shown the promising benefits of ceramides in treating atopic dermatitis.

Topical ceramides have been proven to move into the upper layers of skin, promoting hydration. Natural ceramides, derived from plants like wheat germ, penetrate the skin better than synthetic ceramides, but can also be expensive. You will see them listed on the product labels as "ceramide AP", "ceramide EOP", "ceramide NG", "ceramide NP", "ceramide NS", "phytosphingosine" or "sphingosine."

Another way to get the benefits of ceramides is with one of my favourite DIY treatments, designed to nourish, moisturize and improve skin tone.

Unmask Your Glow

1 tablespoon olive oil mayonnaise
1 teaspoon wheat germ
2 or 3 drops bergamot essential oil

1. Mix the mayonnaise and wheat germ in a small bowl.

2. Add the bergamot oil and mix.

3. Apply this mixture to the face, covering it evenly and completely.

4. Let the mask sit for 15 to 20 minutes.

5. Rinse with lukewarm water.

6. Repeat twice a week in the evening for best results.

Trehalose

Have you ever wondered how a desert plant is able to bloom, or resurrect itself, after months of rain deprivation? Or have you thought about how dried mushrooms magically spring back to life when dropped in water? The answer is trehalose.

This naturally occurring sugar forms a gel as plant cells start to dehydrate, stopping them from falling apart and helping them get back into shape. Trehalose can do the same for your skin.

Science shows that when applied topically, trehalose possesses a massive ability to retain water and can help strengthen a parched and weakened skin barrier. Keeping your skin hydrated is one of the keys to making skin appear smooth and young. While trehalose is not an antioxidant, it does offer protective antioxidant effects. For example, the journal *Biomedical Reports* found that trehalose can help protect the skin against UV damage.

Found in both skincare and haircare products, this targeted topical helps retain moisture and has been shown to reduce odors that stick to your hair and body. Note that this ingredient can also be listed on personal care products as "mycose". For best results, look for brands that contain at least 1.5 per cent trehalose. You can find pure trehalose in stores and online.

You can use trehalose to exfoliate and moisturize dry skin using this facial scrub.

Glow-To Facial Scrub

1 tablespoon trehalose
1 tablespoon tea seed oil
2 or 3 drops bergamot essential oil (optional)

1. Mix the trehalose and tea seed oil in a small bowl. Add the bergamot oil, if desired.
2. Apply the mixture evenly to the face.
3. Gently massage in small circles for 1 minute.
4. Rinse thoroughly with lukewarm water.
5. Repeat as needed two or three times per week.

Tea Seed Oil (Camellia Oil)

Tea seed oil (also known as camellia oil) is a fantastic polyphenol-rich oil that is great for use on your skin. To promote healthy, beautiful and youthful skin, use it as a moisturizer or base for skincare products. The intensity of healing properties present in polyphenols protects skin from the effects of ageing (yes, sun damage), skin diseases and damage from wounds and burns. The oleic fatty acids found in tea seed oil may also be helpful in treating skin conditions such as acne, eczema and dry skin, due to the presence of compounds that strengthen the integrity of cell membranes and promote cellular repair.

Polyphenols

These naturally occurring compounds, found in grape skins, red wine, green tea and blueberries, not only activate autophagy but are also potent antioxidants. This is important because your body uses its own supply of antioxidants to defend against free radicals that affect skin's appearance.

But, as we get older, the Inevitable and Accelerated Agers (see page 32) can leave that antioxidant supply depleted. This can lead to rough, sagging and weak skin. The good news: You can replenish that antioxidant power with potent topical polyphenols, helping to repair damage and restore elasticity to prematurely ageing skin.

A growing body of evidence shows resveratrol and EGCG can both slow skin ageing when applied topically. According to the *Journal of Cosmetic Dermatology,* researchers found that topical resveratrol possesses seventeen-fold greater antioxidant potency than an expensive and powerful drug used in some antiwrinkle creams. The *Archives of Dermatological Research Journal* showed the polyphenolic compounds present in green tea, including EGCG, also provide natural anti-inflammatory effects to the skin. When applied topically, these compounds help address many different inflammatory conditions of the skin, including acne, dermatitis and rosacea.

The *Journal of Dermatologic Therapy* reported that polyphenols not only improved skin elasticity, but also increased moisture level and smoothness when applied topically. Polyphenol-containing skincare products include skin cleansers and face masks, moisturizers and sunscreens. In skincare products, polyphenols will most likely be listed by their individual plant names – all parts of the plant can be used, including leaf, seed, skin, root, seed, shoot, vine, sap, flower and fruit, in multiple forms such as powders, extracts, juices and oils. Note that antioxidants will degrade when they are exposed to light and air, so if your polyphenols are in a jar, once you open it, they won't work as well. Look for airtight, opaque packaging.

The DIY polyphenol treatment overleaf is one of my favourite ways to combat redness – from rashes to blemishes and even eczema.

On-the-Glow Spot Treatment

Green or black tea bags

1. Fill a cup with warm water (a mix of 2 parts hot water to 1 part cold water works best).
2. Dip the tea bag into the water and let it steep for a few minutes. (Polyphenols like the EGCG and bergamot found in the green and Earl Grey teas, respectively, are more potent together, so feel free to combine them for added youth-boosting benefits.)
3. When the water is dark, take the tea bag out and dab some of the water onto any red spots. Repeat this for a few minutes; the tea will dry quickly, so feel free to use a lot of the water.
4. Leave the tea on overnight – don't wash it off right away!
5. In the morning, rinse off any tea residue. You should notice firmer, softer skin and a reduction in your spots and any red marks.

Caffeine

Produced by plants and occurring naturally in coffee and tea, caffeine has the ability to penetrate the skin barrier. It works as an antioxidant and a vasoconstrictor.

As an antioxidant, caffeine can help fight future signs of ageing. When topically applied, it helps protect cells against UV radiation and slows down the process of photoageing of the skin. A study done by the University of Washington and published in the *Journal for Investigative Dermatology* exposed healthy human skin cells and UV-damaged skin cells to caffeine. The caffeine caused the UV-damaged cells to die but did not affect the healthy cells. This may explain caffeine's "sunscreen" effect.

Caffeine also works as a vasoconstrictor, constricting small blood vessels and reducing inflammation. It can be found in a number of creams designed to minimize dark circles and sagging skin under the eyes. In addition, its constricting effect can help reduce redness.

Dry Brushing

This practice involves rubbing a dry brush in a circular motion on your dry skin. It stimulates exfoliation by sloughing off dead skin cells. It also promotes circulation.

A proper dry brush should have firm bristles and a long handle for hard-to-reach areas. You can find them in stores or online for around £10.

Working from the outside in, start at your hands and arms or feet and legs and move toward your heart. The process should take 10 to 15 minutes.

Caffeine derived from coffee beans and tea leaves can be found in a variety of skincare products, from moisturizers to soaps. One of my favourite ways to get the benefits of caffeine is with a DIY scrub. One of the best I've ever found came from Elle.com, which I've adapted here.

Make Cellulite Glow Away

1 cup coffee grounds
3 tablespoons sea salt flakes
6 tablespoons coconut oil
Dry brush (optional)

1. Combine the coffee grounds and salt in a large measuring cup or bowl.
2. If the coconut oil is solid, melt it in the microwave; if liquid, simply pour it into the measuring cup or bowl with the coffee grounds and salt. Mix well.
3. Transfer to a waterproof jar or other airtight container and store it in the shower.
4. If desired, before showering, use a dry brush (see box above) to exfoliate and stimulate blood flow.

5. In the shower, apply the scrub to areas with cellulite. Massage the scrub onto your skin in a circular motion, then rinse.
6. Repeat every time you shower for best results.

Try to use as many of the autophagy-activating topicals – ceramides, trehalose, polyphenols and caffeine – as possible, because a combination is best to outsmart wrinkles and get glowing skin. To be sure you are getting a quality amount of these actives, make sure they appear among the top third of your skincare product's ingredient list.

Treat Yourself to Youth-Boosting Therapies

You already know about the stress associated with fasting and exercise to activate autophagy – but here's one more reason stress doesn't have to stink. Therapeutic mechanical stress – causing minor trauma to cells in skin and muscles – may elevate autophagy and execute repair. That means that manipulating your skin through therapies like massage and acupressure can stimulate your cellular cleansing cleanup crew to help you look and feel younger. What a perfect excuse for a spa day!

Try one of these youth-boosting treatments to activate mechanical stress, making your cells act younger.

Facials: A great way to remove dead skin cells and stimulate cell turnover, facials can help to smooth skin and boost a youthful glow.

One of the most promising facial treatments for autophagy activation and youthful-looking skin is microneedling, a cosmetic procedure that involves repeatedly puncturing the skin with tiny sterile needles that range between 1 millimetre and 1.5 millimetres in length. Being pricked over and over again with tiny needles may seem like a form of torture, but I can tell you from firsthand experience, it may be one of the best things you can do to rejuvenate your skin.

Many dermatologists believe microneedling may be an anti-ageing breakthrough – and it works by creating a stress-response mode in your body. The theory is that the needles stress your cells to activate autophagy and repair the damage. The procedure boosts production of collagen and elastin for an anti-ageing effect. A study in the *International Journal of Dermatology* found it to be a promising, minimally invasive treatment option to help advance collagen production. It can be effective for dark spots, redness, and fine lines and wrinkles. The procedure is most commonly performed by a doctor or trained beautician, but at-home versions also exist and are gaining popularity.

Massage: This therapy dates back more than 5,000 years. The benefits of massage have been well examined, proven, and recorded throughout its history.

Archaeological discoveries indicate that prehistoric people used herbs on their bodies to promote health and healing. Chinese literature records that massage was used for healing as far back as 3000 BCE. It is believed that Hippocrates, the Greek physician, was the first to use massage for circulation, maintaining that the strokes should be made in the direction of the heart. And during the First World War, massage was used along with surgical treatment to alleviate pain, reduce oedema, assist circulation and promote tissues nutrition.

Today massage is more than just a luxurious form of pampering. It's been shown to help reduce anxiety, aches and pains, and relieve sore muscles. Its benefits are so widespread that it's offered everywhere, including doctors' offices, spas and even airports.

There are many different types of massage. Some of the most common therapies include:

- Swedish massage, a gentle form of massage that uses long strokes, kneading, deep circular movements, vibration and tapping to help relax and energize you.
- Deep-tissue massage, a technique that uses slower, more forceful strokes to target the deeper layers of muscle and connective tissue, commonly to help with muscle damage from injuries.
- Sports massage, similar to Swedish massage, but geared toward people involved in sports activities to help prevent or treat injuries.
- Trigger-point massage, which focuses on areas of tight muscle fibres that can form in your muscles after injuries or overuse.

Make sure your massage therapist is registered with the Complementary

& Natural Healthcare Council (CNHC), or belongs to a professional body recognized by the CNHC.

Acupuncture: This popular alternative treatment originates from traditional Chinese medicine. Practitioners use needles to trigger specific points throughout the body to release the flow of energy.

Acupuncture has been shown to benefit a variety of ailments from digestive issues to neurological problems, sinusitis to mood disorders.

Ten to twelve needles are inserted by hand. The areas of insertion correlate to your specific ailment. The most important preparation you can do for this treatment is to find a certified or licensed practitioner. I found acupuncture to be painless.

Acupressure: This traditional Chinese treatment is known as needleless acupuncture. Hand or elbow pressure is generally used on the same trigger points as acupuncture. The goal again is to clear the flow of energy in the body. The major difference between this and acupuncture is that acupressure can be self-administered. (See page 143 for a DIY guide to acupressure to help your skin glow.)

By stimulating specific points, you have the ability to alleviate tension, increase circulation and reduce pain and other ailments for well-being.

Facial acupressure treatment is believed to improve the condition and tone of muscles and connective tissue, which can lessen the appearance of wrinkles. It can also help in weight-loss maintenance.

Facials, massage, acupuncture and acupressure – along with other treatments that manipulate the body to cause minor trauma to cells – can help to stimulate autophagy. While some of these therapies can be practiced on your own or with a partner, if a treatment involves deep muscle and tissue manipulation, it is best to see a licensed therapist. I try to do one of these treatments each month to help change the way my skin ages.

You now have the tools to glow younger – from beauty to sleep to exercise to nutritional supplements and diet. In just fifteen days, you should see noticeable changes in your weight, your energy, your mood and your skin.

I believe success is where opportunity and preparation meet, and I want you to succeed. Now is your opportunity, and in the next chapter, I'll prepare you with everything you need to get glowing.

DIY Acupressure

Acupressure points to help activate autophagy also work to boost your glow.

The following is a guide to the most commonly used acupressure points for alleviating skin conditions.

For each, use your index or pointer fingers to apply pressure for 30 to 60 seconds. Repeat three times per week.

Third Eye Point

WHERE IT IS Directly between your eyebrows at the point where the bridge of your nose meets the centre of your forehead.

WHAT IT DOES Stimulates the main endocrine gland, the pituitary, helping to enhance the overall appearance of your skin.

Four Whites

WHERE IT IS Near the top of your cheekbone, about a finger's width below your the eye socket in line with the centre of your iris.

WHAT IT DOES Helps alleviate acne and facial blemishes.

Facial Beauty

WHERE IT IS At the bottom of the cheekbone, about a finger's length directly below your pupil.

WHAT IT DOES Improves circulation, swelling and blemishes.

Wind Screen

WHERE IT IS In the indentation directly behind your ear lobe. (This is easiest to feel if you open your mouth and feel for the indentation between your ear and jaw.)

WHAT IT DOES Believed to help balance your thyroid gland and increase skin's radiance; may also help to relieve hives.

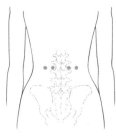

Sea of Vitality

WHERE IT IS In your lower back (between the second and third lumbar vertebrae), two to four finger widths away from your spine at waist level.

WHAT IT DOES Fortifies your immunity, helping to relieve redness, eczema, and bruises on the body.

GLOW15 SUCCESS STORY

It's like Glow15 helped me get younger!"

IOLA

"I'm in that gorgeous state of menopause and I just wanted to get back to feeling like my best me. Glow15 made that possible. **It changed my mindset. I just became more positive. I worked muscles I forgot I had and I slept better than I have in years, which meant I woke up happier.**

"But the biggest change may have been to my skin – I added products that had the autophagy-activating ingredients **And instead of the dullness and dryness I'd had since menopause, my skin is bright and smooth. It's like Glow15 helped me get younger!"**

Chapter 9

Know Before You Glow

How to Prepare for Glow15

"By failing to prepare, you are preparing to fail." —
BENJAMIN FRANKLIN

At this point, I've laid out the foundation of Glow15, and if I've done my part correctly, you should have a basic understanding and be excited about autophagy and how to make it work for you. Hopefully, you've also been able to follow my plan of what to do, when to do it and why it's important. Now, before you begin to elevate your life, I want you to be prepared for success. To properly prepare, you need to do so mentally and technically.

Mental prep is most important. This is where you need to be bold. You need to think of yourself as a priority. You need to believe (or at least pretend until you believe) that you deserve it. Glow15 can give you the tools to look younger and feel better, but only you can give yourself permission to use them. Before you do any other preparation for this plan, tell yourself that you are entitled to do this. Give yourself consent to succeed. Now, what does success look like for you? Write it down. Yes, I'm serious. I read a startling statistic that 95 per cent of people who don't have written goals fail, and the 5 per cent who do have written goals succeed. So write it down. And start looking forward to seeing your own success.

Technical prep is what gives you the readiness to achieve your goals. These are the tools that ensure you don't just survive, but thrive. Whether you want to look younger, feel better, lose weight, gain energy – however you want to boost your life, I want to help you succeed. And here I've outlined (1) the guidelines, (2) the checklist and (3) the metrics to help you technically prepare to get glowing.

The 15 Guidelines to Get Glowing

These are the major tenets of Glow15, all in one place and simplified.

1. Practice IFPC. Glow15 cycles you through High and Low days. You'll practice intermittent fasting (periods of unrestricted and restricted eating) along with protein cycling (periods of normal/high protein consumption and low protein consumption).

2. Fat first, carbs last. You'll start every day with fat – eating an AvocaGlow or Egg15 – to promote autophagy. Eating fat as your first meal can regulate your metabolism and keep you satiated throughout the day. You'll save your carbs for your last meals. This will help your body process fat as fuel throughout the day.

3. Drink AutophaTea. My ultimate youth boost. This tea, made of autophagy-activating ingredients, including whole citrus bergamot Earl Grey tea, green tea, cinnamon and coconut oil, can be enjoyed on both High and Low days. (You'll find the recipe on page 199.)

4. Eat autophagy-activating foods. Include a variety of nutrients — polyphenols, sphingolipids, omega-3s, sulforaphanes, vitamins C, D, E and K, spermidine, saponins and probiotics — to help your cells clean up and repair themselves.

5. Protect and repair with Powerphenols. This is your nutritional insurance policy. The highly potent actives in these nutritional supplements not only help defend your cells from disease, but also work to mend existing damage.

6. Try the fab four. Resveratrol, curcumin, EGCG and berberine are the nutritional supplements that best activate autophagy and promote youthful cells. Use them with your diet and rejuvenate your body.

7. Exercise on High days for 30 minutes. This means you'll work out four times per week for a minimum weekly total of 120 minutes.

8. Do HIIT and RET workouts. High-intensity interval training (HIIT) and resistance exercise training (RET) are the two best workouts for activating autophagy. You will do each twice a week.

9. Run on empty, pregame with caffeine, refuel with protein. Don't exercise on a full stomach, but drink some caffeine before your workout. Afterward, eat some protein.

10. Keep it hot and keep it going. Heat can turn up the spark on your autophagy. And research has shown that you can lose that spark if you stop working out for more than two weeks, so move it to keep it!

11. Tailor sleep to your bird type. Are you a Lark, a Hummingbird or an Owl? Knowing your bird type can help you get a better night's rest. Take the quiz on page 107 to determine your bird type.

147

12. Customi-Zzz. Once you determine your bird type, find out the best way to get some rest. Discover how to let in light and set your bedtime.

13. Power down, cool down, null the noise. For all birds who want to get better sleep and wake up energized and refreshed, practice good sleep hygiene. Before bed, turn off electronics, set your thermostat to cool and play coloured sound (see page 119).

14. Targeted topicals. Try ceramides, polyphenols, trehalose and caffeine on your skin to help make it glow.

15. Treat yourself! From DIY beauty treatments to spa therapy, wine to chocolate, remember that Glow15 is about making yourself a priority. Be bold enough to put yourself first.

The Checklist to Get Glowing

I love a list! I make lists for everything, from my daily to-do list to travel and packing lists to my children's needs-and-wants list to gift lists and thank-you lists – you name it, I've got it on a list.

And I've made a list for you, too. It includes all the essential needs for Glow15. First, use this comprehensive checklist to mark off what you already have in your store cupboard, fridge, medicine cabinet, beauty drawers and elsewhere in your home. If you're missing an item, grab it from your local supermarket, neighbourhood superstore or online.

Here are some shopping guidelines – read these over before you head to the shops with your list.

Glow15
Shopping Guidelines

- Whenever possible, opt for organic produce – you'll be avoiding unnecessary chemicals in addition to getting the most nutrients possible.

- Frozen organic fruits and vegetables are a lot more affordable and just as nutritious! They're a great substitution for any fruit or cooked vegetable in a smoothie.

- Whenever possible, opt for grass-fed beef; organic, free-range poultry; organic pork; and wild-caught, sustainable fish. Not only does this support the environment, it also gives you a better nutrition profile that can help with autophagy activation.

- Whenever possible, opt for organic dairy products like butter, yogurt and ghee (clarified butter). Again, this provides more nutrients and saves you from unnecessary added hormones, antibiotics and synthetic toxic input.

- Choose pink Himalayan salt for the added mineral traces it provides (but remember an excess of any type of salt – high in sodium – is bad).

- As much as you can, buy fresh food over canned. The few exceptions to this are canned tomatoes, canned beans and canned salmon (look for a wild-caught, sustainable brand). The cans should be BPA-free, if possible.

- Opt for cold-pressed, unrefined oils when you can.

Pantry

- Almond flour
- Apple cider vinegar, raw
- Avocado oil
- Beef stock
- Black beans
- Black olives, pitted
- Canned tomatoes (with skins, crushed)
- Capers
- Chia seeds
- Chickpeas
- Coconut oil
- Coffee, ground
- Dark chocolate (80% cacao)
- Dijon mustard
- Dulse flakes
- Extra-virgin olive oil
- Green olives
- Green tea
- Lentil pasta
- Macadamia nuts
- Maple syrup, pure
- Mayonnaise, avocado-based
- Mirin
- Miso paste
- Nut butter, peanut or almond, unsweetened
- Plain chocolate chips
- Quinoa
- Raw honey
- Rice (brown or black)
- Rice cakes, whole-grain
- Sesame seeds
- Sherry vinegar
- Sun-dried tomatoes
- Sunflower seeds
- Tahini
- Tamari, gluten-free, or coconut aminos
- Toasted sesame oil
- Tomato salsa
- Unsweetened coconut
- Vegetable bouillon (look for a brand without "yeast extract" in the ingredients)
- Walnuts
- Whole citrus bergamot Earl Grey tea
- Wheat germ
- Wooden skewers

Herbs/Spices

- Basil, fresh and dried
- Bay leaf
- Black pepper
- Cayenne pepper
- Chilli flakes
- Cinnamon, ground and sticks (true Ceylon cinnamon, not cassia or Chinese cinnamon)
- Chives, fresh
- Coriander, fresh
- Cumin, ground
- Curry powder
- Dill, fresh and dried
- Garlic, fresh and powder
- Ginger, fresh and ground
- Mint, fresh
- Oregano, fresh and dried
- Parsley, fresh
- Rosemary, fresh
- Sea salt flakes
- Thyme, fresh and dried
- Turmeric, ground

Proteins

- Bacon (nitrate-free)
- Beef, boneless chuck steak
- Beef mince
- Chicken thighs
- Eggs, large
- Ham, sliced
- Pork loin roast
- Prawns
- Sausage
- Salmon, cold-smoked
- Salmon, wild, fillet and canned
- Turkey, mince and sliced roast turkey breast

Fruit

- Apples
- Avocados
- Berries (blackberries, blueberries, strawberries)
- Grapefruit
- Lemons
- Limes
- Tomatoes and cherry tomatoes

Vegetables

- Asparagus
- Artichoke hearts
- Broccoli
- Broccoli sprouts
- Brussels sprouts
- Cabbage, red
- Carrots
- Cauliflower
- Celery
- Chard
- Courgettes
- Cucumbers
- Jalapeño chillies
- Kale
- Kimchi
- Lettuce, romaine, Little Gem, baby leaves
- Mushrooms, fresh, shiitake, portobello and chestnut
- Onions
- Peppers
- Rocket
- Sauerkraut
- Spaghetti squash
- Spinach
- Spring onions
- Sweet potatoes

Dairy/Dairy Alternatives

- Almond milk, unsweetened
- Butter or ghee (clarified butter)
- Cheddar cheese
- Coconut milk, full-fat
- Feta cheese, full-fat
- Goats' cheese, full-fat
- Greek yogurt
- Mozzarella cheese, full-fat
- Pecorino Romano cheese

Alcohol

- Wine
- Beer

Powerphenols

- Resveratrol-trans (250-mg capsules)
- Organic curcumin (500-mg capsules)
- Berberine (500-mg capsules)
- EGCG (300-mg capsules)

Beauty Tools

- Epsom salts
- Dry brush
- Tea seed oil (camellia oil)
- Trehalose

The Metrics to Get Glowing

While success should really be measured in the way you feel, your self-esteem and your confidence, one of the most motivating ways to reach your goals is to have a written account of your journey. For many of you, this may mean keeping a diary of your experience. But I need to see the numbers, and I think that you will want to see yours. While you can expect changes to both how you look and how you feel in just fifteen days, Glow15 is not a temporary plan. Glow15 is a permanent lifestyle, and you should expect to notice even more differences during the next fifteen days, and more on the days after that, and the ones following those and so on.

Below, you will find the metrics to measure your success along the way. I've described how to calculate each and provided a worksheet to take you through measuring your success on your first year of Glow15.

Goal: The first and most important metric is setting a goal. If you read the beginning of this chapter, you should already have this written down. Now put it at the top of your worksheet. Again, I want you to achieve success – and all successful endeavors start with goal setting.

Weight: Write down your current weight. If you have bathroom scales, weigh yourself on them, or use the scales at the gym or your doctor's office. Weight is not always a precise form of measurement because it fluctuates and doesn't reflect the ratio of lean mass to fat mass in your body, but it is still a good baseline. After fifteen days on Glow15, weigh yourself again. It's best to do it first thing in the morning before you eat, but just make sure you weigh yourself at the same time. I personally really like to use digital scales like the Nokia Body (or Body+) scales. They're expensive, but offer a full-body composition breakdown, which can give you a more detailed assessment of your changes.

A note about weight: you will likely gain lean muscle on Glow15. Lean muscle actually helps speed up your metabolism, but that muscle has mass that can show on the numbers on your scale. I want you to keep this in mind so you don't let the scale define you. Yes, some women on Glow15 lost over seven pounds in the first fifteen days, but others lost inches and dropped a dress size while never seeing a big change in their weight.

Body Mass Index (BMI): This is the ratio of your height to your weight; you can find many online BMI calculators. A BMI of 25 or over is considered overweight. Doctors often use BMI as a marker for health, because this measurement takes into consideration not just your weight in absolute terms, but in proportion to how tall you are. This is also not a perfect measure, because it doesn't take into account lean muscle mass, but it can give you some clues about changes in your body over time. With each pound of weight you drop, your BMI will change.

Waist Size: Measure your waist right at the belly button. You can also take measurements of whatever else you like – your thighs or arms, for instance. Just measure the same place and the same way each time. It's very likely you will see changes in your shape before you drop any weight. That's an indication that your lean muscle versus fat mass is changing.

Clothing Size: Write down your current clothing size. If you're anything like me, there are a variety of different sizes hanging in your wardrobe. You may have your "skinny-day" jeans, your "fat-day" jeans, your "everyday" jeans and your "never but can't get rid of them" jeans. After your first fifteen days on Glow15, check your size again – either from your own wardrobe or on a shopping excursion (I won't tell you to buy, but you can still try). Allow yourself to notice any differences. Does the same size fit differently? Have you changed sizes? Has your body composition changed, making the clothes look different on your body? Some of this may be subjective, but it's one of my favourite ways to measure progress.

Blood Pressure: If you have your own blood pressure cuff, take your pressure first thing in the morning before eating or drinking. If you don't have one, many pharmacies offer free blood pressure screenings. Don't take your blood pressure after having caffeine or exercising, as both of these can cause a temporary rise. This is one of the best ways to gauge improvement in your health, because the higher your blood pressure, the more strain on your circulatory system. Normal blood pressure is considered 120/80 or lower.

Resting Heart Rate: To determine your resting heart rate, the NHS website suggests you find your pulse on your wrist, count the number

GLOW15 SUCCESS STORY

"I like myself more now, and I'm more hopeful than I've ever been about love"

SARAH

"I've always been a little heavier than I should be – plus, I'm over 30 and a single mum – that in itself makes dating a struggle. Then, I'd compare myself to all the younger, thinner, fitter women and I felt like I didn't measure up. So I gave up. I gave up the singles scene, I gave up on the online matchmaking sites, I gave up flirting, and I gave up trying to date. I was about to give up on myself when I was convinced to try Glow15.

"I really appreciated how the plan was laid out – I not only found what to eat, but when to eat. Same thing with exercise – it outlined not only what to do, but when to do it, and why. I understood and appreciated the principles behind the plan. And I liked how **I could easily make it work with my busy life. And for the first time in a long time, I liked my measurements.** I didn't even lose a lot of weight, but I did lose inches: **3 inches off my thighs, 4 inches in my shoulders, and more all over – my chest, my waist, my hips. Glow15 gave me a confidence I didn't think was possible,** let alone in only 15 days. I like myself more now, and I'm more hopeful than I've ever been about love."

of beats for 30 seconds, and then multiply that by two. In general, the lower, the better. That's a sign that your heart has to do less work – and that's a good thing. Take it when you first wake up, while you are still lying in bed.

Track Your Glow

Use the worksheet on page 158 to chart your own progress and see the amazing results. Of course, you can include more benchmark periods if you like, such as at the start of every month.

Optional: Blood Test

If you haven't had your blood checked, it's a good idea to get some baseline numbers, such as those in the chart that follows. Check with your doctor, as he or she may want you to fast prior to having your blood drawn. Your doctor will report whether your numbers are normal, high or low in each area. You certainly won't be able to get blood tests every week or month, so ask your doctor if you can have one at the six-month mark.

The combination of mental and technical preparation is key to your success on Glow15 because it encourages you to think about your actions. Thinking through what you will do and when you will do it is important to understanding what it will mean for you to accomplish your goals.

Proper prior planning enables you to identify potential obstacles. What parts of the plan are going to be more difficult for you? Identifying the potential obstacles now – which is only possible through preparation and planning – gives you every advantage for conquering hurdles later. And the opposite is true as well. Planning enables you to identify your strengths and potential opportunities to increase your likelihood of success.

I know you can succeed on Glow15. I believe you will reach your goals. To that end, I want to give you every chance to triumph. To help you further prepare, in the next chapter, I will show how you can get cooking with Glow15 and the best ways to prepare your kitchen for a youth boost. Following that, I've laid out a detailed sample fifteen-day plan to help you get glowing.

At-Home Self-Tests

GOAL								
	DAY 1	DAY 15	DAY 30	DAY 60	MONTH 3	MONTH 6	MONTH 9	1 YEAR
BMI (Body Mass Index)								
Waist Size								
Body Part 2 Size (your choice, like hips or thighs)								
Body Part 3 Size (your choice, like hips or thighs)								
Clothing Size								
Blood Pressure								
Resting Heart Rate								

Blood Work

	DAY 1	MONTH 6	1 YEAR
Total Cholesterol			
HDL ("good") Cholesterol			
LDL ("bad") Cholesterol			
Fasting Blood Glucose			
Triglycerides			

Chapter 10

Get Cooking with Glow15

The Youth Boost in Your Kitchen

Preparing meals on any new plan can be intimidating. It doesn't matter if you are a master chef or have not yet mastered boiling water — I get it. How you cook is as important as what you cook. In this chapter, I'll spill the beans on the best ways to cook to activate your autophagy, from what to do and when to do it to which tools to use and why it matters. I'll show you exactly how to cook to maximize your youth.

From Soup to Nuts

Before you start making the Glow15 recipes in chapter 12 or creating your own meals to get glowing, it's important to understand how different cooking methods can affect the autophagy-activating nutrients in your food. You already know the benefits of cooking — it can concentrate tastes and flavours, soften tough foods and break starch molecules into digestible fragments. But cooking can also create dangerous chemical reactions in sugars, amino acids (the building blocks of protein) and creatine (a chemical found in muscle and also in the brain), producing compounds that damage your cells. Some techniques inadvertently can also damage proteins and oxidize fats.

This negative impact is most likely to happen when your food is cooked at high temperatures, which can make healthy foods suddenly unhealthy in your body. My general cellular youth-boosting rule is to choose shorter cooking times and lower cooking temperatures, as this will help ensure you maintain the quality of the nutrients you ingest.

To make things easy, I divided the Glow15 cooking methods into two categories: Cream of the Crop, the best ways to cook for autophagy activation, and Bottom of the Barrel, the methods you should use sparingly, as they can inhibit autophagy.

Cream of the Crop

These are my top cooking techniques. They're the best for retaining maximum nutrients – plus, they're simple and efficient.

Steam: Steaming your vegetables is a wonderful way to soften tough outer skin or plant cellulose while still retaining nutrients. Most green vegetables take only a few minutes to steam; starchy vegetables like sweet potato naturally take longer. Be careful not to overcook (or burn yourself – we sometimes forget how hot steam really is!); you want to cook the vegetable until it is tender but still has a brilliant, bright colour. You know you have overcooked (and lost nutrients) when your broccoli or green beans turn a drab army green.

To steam vegetables, cut them into pieces of roughly the same size so they will cook uniformly. Place a steamer basket into a large pot and fill the pot with 1 to 2 inches of water – the water level should be below the bottom of the steamer basket. Bring the water to the boil over high heat. Add the vegetables, cover and reduce the heat to medium. Cooking time will depend on the type and size of the vegetable. Carefully check for doneness after 2 to 3 minutes (and set a timer if you get easily distracted!). Vegetables are done when they are tender and can easily be pierced with a fork.

Poach: Poaching is another healthy choice when it comes to maintaining nutrients and is best used for delicate foods like eggs, fish, chicken breasts and fruit like pears. When poaching, the food cooks in a liquid, such as water, stock or wine, heated to just below a simmer. Poaching is an ideal way to cook eggs, as you can cook the white but leave the yolk runny and uncooked (but heated), keeping the vital enzymes, cholesterol and lipids intact and protected from oxidation. Note the difference between poaching and boiling: poaching is cooking at a relatively low temperature, whereas boiling cooks food at a much higher temperature. Nutrient loss can easily occur through evaporation when a food is boiled.

To poach an egg, bring a pot of water to a boil and add a splash of distilled white vinegar to the water (this helps the proteins in the egg white coagulate and stay together). Reduce the heat to low to keep the water at a low simmer. Crack an egg into a small dish. Using the handle of a wooden spoon, stir clockwise to create a gentle whirlpool in the water. Slowly pour the egg into the swirling water. Cook for about 3 minutes, until the egg white is cooked through but the yolk remains runny and soft. Remove with a slotted spoon to drain excess water. You can poach more than one egg at a time in the same pan, just be sure to crack them into the small dish and add them to the water individually.

Sauté: Sautéing is a quick cooking method that uses a small amount of fat or oil in a shallow pan over relatively high heat. Vegetables will still be firm after cooking and nutrients are retained because the cooking time is fairly short.

To sauté mushrooms, remove the stems and chop the mushroom caps into bite-size pieces. Heat a medium frying pan over medium-high heat. When the pan is hot (this won't take long), add avocado oil and let it heat for 1 minute. Add the mushrooms, sprinkle with sea salt flakes and cook, stirring continuously to prevent burning, until the mushrooms start to brown and stop giving off liquid.

Bake: Baking uses dry heat to cook foods for fairly long periods of time. A variety of foods can be baked, from breads and desserts to meats, fish, and starchy vegetables.

To bake salmon, preheat the oven to 180°C /Gas Mark 4. Line a baking tray with aluminium foil or nonstick baking paper. Place salmon fillets (use centre cuts for optimal results and uniform cooking) skin-side down on the prepared tray. Drizzle with avocado oil and sprinkle with sea salt flakes and freshly ground black pepper. Bake for about 20 minutes, until just cooked throughout. Leave to rest for 2 minutes and serve immediately.

Bottom of the Barrel

These are the techniques I limit or avoid altogether when I can. Not only can they harm the food and weaken its nutritional value, but they can be messy and labour-intensive.

Grill or Barbecue: A summertime favourite, grilling is best done minimally because it uses high temperatures to cook meat and, less often, vegetables and fruit. Grilled meat has a golden to dark brown colour and distinctive aroma and flavour from a chemical process known as the Maillard reaction. While you may enjoy the taste, grilling at high temperatures can produce carcinogenic heterocyclic amines (HCAs), which contribute to cellular ageing and hinder autophagy.

If you do choose to grill, do so at lower temperatures to produce fewer HCAs, or try indirect grilling by placing the food to the side of the heat source, not directly above or below it.

Blacken or Char: This method of cooking meat oxidizes the fat molecules, which inhibits the process of autophagy. Oxidizing fat molecules makes them inflammatory and can also cause hormone disruptions that can lead to weight gain.

Deep-Fry: Heavily fried foods are another food to minimize due to high temperatures and oxidation. Typically, low-quality oils are used repeatedly to cook foods, leading to rancid oxidized oil, contributing to inflammation and the formation of free radicals in your body, and creating obvious obstacles for autophagy.

Icing on the Cake

Cooking autophagy-activating foods will provide nutrient-dense fuel for your body, focused energy for your mind, and ideal nutrients for your cells. If you follow the Glow15 guidelines, you will experience those benefits. But do you want to achieve even greater results? You can – and it's as easy as eating apple pie! Here, I'll go through the tools you can use when cooking to boost your glow.

A Fat Lot of Good: Choose the Right Cooking Oil

As you already know, fat is a key component of Glow15 – and one of the best ways to get fat is through healthy oils. Oils can add satisfying richness and flavour to food, help absorb the health-giving properties of herbs and spices, and aid in Glow15 cooking. And whether you drizzle it, splash it or pour it on, preparing food with healthy oils can provide an array of autophagy-inducing benefits.

But not all oils are created equal. The key to your body thriving is knowing when to use what. Factors like smoke point, fatty acid profile, and taste all matter when cooking with oil.

Smoke Point: This is the temperature at which oil begins to smoke or burn. Different oils withstand varying levels of heat. When an oil is heated to its smoke point, the flavour changes, the oil loses its nutritional value and, even more frightening, it produces toxic fumes. When you breathe in or consume the oil, it causes damage to your cells and can potentially lead to disease.

The Cleveland Clinic categorizes oils as high, medium-high, medium, and no-heat. Oils with a high smoke point are best for searing, frying, browning, baking, and oven cooking. The best example I've found is avocado oil, which has a smoke point of 260°C. Oils with a medium-high smoke point are best suited for baking and cooking. Extra-virgin olive oil, with a smoke point of 190°C, falls into this category. Medium-smoke-point oils are best for light sautéing, using in sauces and low-heat baking. Coconut oil has a medium smoke point of 175°C. No-heat oils should not be heated at all and are best used in dips, dressings and marinades.

Flaxseed is an example of a no-heat oil.

Fatty Acid Oxidation: The most important factor in the fatty acid profile of an oil is how susceptible it is to oxidation.

Oils can be saturated, unsaturated and polyunsaturated. The less saturated the oil is, the more vulnerable it is to damage by heat. Saturation creates a barrier against light and heat. Saturated fats are saturated with hydrogen atoms, creating the strongest barrier and making them the least susceptible to oxidation, whereas monounsaturated fats are missing one pair of hydrogen atoms, making the barrier a little weaker, and polyunsaturated fats are delicate, missing multiple pairs of hydrogen atoms, making them the most susceptible to oxidation.

If you take a look at an oil high in saturated fat, like coconut oil, you will see that it's low in polyunsaturated and monounsaturated fat. Its profile is 2 per cent polyunsaturated, 6 per cent monounsaturated and 87 per cent saturated. This makes it good for cooking at higher temperatures.

On the opposite end of the spectrum are oils higher in polyunsaturated fatty acids. Oils high in polyunsaturated fats are safe for cooking only if they also have strong antioxidant protection. Sesame oil is an example, as it contains sesamolin, which when heated converts to the antioxidant sesamol, providing powerful protection against oxidation.

Oils high in monounsaturated fatty acids are less susceptible to oxidation than polyunsaturated fatty acids. Monounsaturated fats, especially those that are also high in saturated fats, can better withstand low to medium temperatures. Avocado oil has a fat profile of around 8 per cent polyunsaturated, 82 per cent monounsaturated and 10 per cent saturated. It also has a high concentration of antioxidants, as well as vitamin E, or tocotrienols.

Taste: The flavour of cooking oils is varied – some can be strong enough to change the taste of a dish, while others are milder and can blend with your other ingredients. This is why it's important to consider the taste of your oil when cooking. Coconut oil tends to have a tropical scent and flavour, depending on the type you get; those labeled "virgin" and "expeller-pressed" typically have the strongest flavour and scent. Olive oil is fruity and aromatic and is sometimes described as "peppery." Toasted sesame oil has a nutty, rich aroma and is the focus of many Asian dishes. Avocado oil has a light flavour that will enhance rather than overpower a dish.

Strike While the Oil's Hot: Avocado

From smoke point to fat oxidation to taste, avocado oil comes out on top.

The avocado mostly gets attention as the main ingredient in guacamole or in the form of avocado toast — as an oil, it doesn't make the headlines the way olive oil does. But this tasty oil, which is made by pressing the pulp of the avocado, is what you can use in your salad dressings and other recipes. It's loaded with nutrients that initiate autophagy, making you look and feel younger.

Cultivated as early as 500 BC in Mexico, the avocado has a nutritional makeup that helps promote heart health and enhances the action of other vitamins in the body. The oil consists of about 75 per cent monounsaturated fatty acids, with the remainder split between polyunsaturated fatty acids and saturated fatty acids.

Avocado oil is a force of nutrition. It's made up of around 85 per cent unsaturated fat, which has been linked to lower LDL cholesterol levels and harmful triglycerides in the body. Of this unsaturated fat, oleic acid makes up a significant part. Oleic acid has been linked to lower blood glucose levels and improved bloodflow. In addition, it contains lutein, which has been shown to improve eye health and can reduce the risk of age-related eye problems.

Avocado oil supercharges your nutrition. Your body sometimes needs help absorbing various vitamins and nutrients. Avocado oil can help improve that process. One study found that avocado oil helped the body absorb carotenoids from carrots and spinach when eaten on a salad. The increase in absorption was up to 17-fold!

Avocado oil is a protector against ageing. Studies have shown that avocado oil can help protect the body against ageing by fighting damaging free radicals — one of the major villains in the ageing process.

Avocado oil has a mild, understated flavor. It's a pale amber-green color, and has a sweet and pleasant aroma—which I consider my aromatherapy when cooking! It's especially nice for salads because of the avocado taste, but it truly does have a phenomenal culinary versatility.

Its subtle flavours make avocado oil great to use raw, but it also performs well at high temperatures, both because of its abundance of antioxidants as well as its high smoke point. You can learn how to use avocado oil in dishes from stir-fries to salad dressings in the Glow15 recipes starting on page 198.

What to Know Before Choosing an Oil to Glow

Look for cold-pressed, unrefined oils. These labels ensure the oil has been extracted by mechanical means, not using heat or chemicals. Expeller pressing is a healthier way to process oil. It does not involve chemicals but instead uses compression or friction to manually extract the oil. If this process is done in cool, controlled temperatures, the oil is called "cold-pressed".

Most inexpensive cooking oils are extracted using harsh chemicals and high heat. Not only does this change the taste of the oil, but the heat alters the nutritional content and may cause the oil to become harmful.

Virtually all vegetable oils – especially rapeseed, corn and soya bean oils – are processed using chemicals and high heat. Chemically derived vegetable oils in theory are "cleaner" and may have higher smoke pints but the refining process can deplete antioxidants and other valuable plant chemicals.

Choose organic when possible. The quality of the original source – whether it's a nut or seed or anything else—correlates with the quality of its oil. I always choose organic because I believe it is superior nutritionally, in particular when it comes to levels of beneficial phytochemicals like antioxidant and natural anti-inflammatory compounds.

Choose oils sold in dark glass bottles and store them in a cool, dry place. Glass is a more inert material than plastic and it is more eco friendly too, provided you recycle. I also buy only oils packaged in dark glass containers as light can cause sensitive oils to oxidize. To avoid any potential oxidation, I often choose my container from the back of the

Oil	Smoke Point (max value)	Fatty Acid Profile (approx. %)
Avocado Oil	260°C	71% monounsaturated, 13% polyunsaturated, 16% saturated
Rapeseed Oil	204°C	62% monounsaturated, 31% polyunsaturated, 7% saturated
Corn Oil	227°C	30% monounsaturated, 51% polyunsaturated, 14% saturated
Extra-Virgin Olive Oil	190°C	73% monounsaturated, 8% polyunsaturated, 14% saturated
Coconut Oil	175°C	2% monounsaturated, 6% polyunsaturated, 87% saturated
Sesame Oil	175°C	38% monounsaturated, 43% polyunsaturated, 14% saturated
Walnut Oil	160°C	17% monounsaturated, 70% polyunsaturated, 9% saturated

Health Components	How to Use It
High polyphenol antioxidants can lower blood pressure. Almost 90% unsaturated fat to help lower cholesterol. Antimicrobial properties boost immunity.	Ideal for a variety of cooking applications due to high smoke point. This oil can also be used in its raw form and is a great addition to your daily AutophaTea (page 199), too.
Usually processed with chemicals and high heat, but you can also find cold-pressed rapeseed oil in the UK.	Use cold-pressed rapeseed oil in dressings and dips. Also ideal for baking and frying due to low smoke point.
It is high in omega-6 pro-inflammatory fatty acids that have been linked to heart disease and cancers.	This oil should be avoided.
Rich in monounsaturated fatty acids, mostly oleic and oleocanthal, which help protect against inflammation, a key driver of heart disease.	Use for baking, oven cooking or light stir-frying, as well as for salad dressings, dips and in its raw form.
Rich in MCTs, caprylic acid, lauric acid and capric acid, which help to burn fat, build muscle and fight bacteria and fungus like candida.	Use for baking, sautéing, soup and, of course, AutophaTea (page 199).
Contains a wide range of polyunsaturated fatty acids, including sesamol and sesamin, which promote heart health. In addition, contains an organic compound called phytate, which has been directly linked to a reduction in the risk of developing cancer.	Use for baking, oven cooking and stir-frying.
Consumption of walnut oil has been shown to lower total cholesterol as well as LDL ("bad") cholesterol and improve the ratio of LDL to HDL ("good") cholesterol.	Despite the high smoke point, this oil is also high in polyunsaturated acids, making it susceptible to oxidation. For this reason, it is best to use walnut oil in its raw form.

shelf, where it will have had the least exposure to light. To further prevent oxidation and rancidity, it might be best to refrigerate your oils, especially if you live in a warm climate or the temperature in your kitchen fluctuates significantly. Don't worry if the oil solidifies in the refrigerator; it will return to a liquid state as it comes to room temperature.

Sowing the Seeds of Youth: Soaking, Sprouting, and Fermenting

Nature does a brilliant job of packaging a plant's next generation into a tiny, easily transportable seed, filled with all the nutrients it needs to start growing. These nutrients can benefit you, too, helping you look and feel younger. But when ingested, seeds can be selfish, keeping their nutrients intact rather than releasing the healthy actives that help promote autophagy. Not to worry – you *can* get seeds to share. The solution? Soaking, sprouting and fermenting.

Soaking softens the tough, outer part of the seed and acts as a predigestion step, making our digestive tract work less to extract the beneficial fats, proteins and nutrients from the nut or seed.

Here is a basic recipe to follow when soaking most varieties of nuts and seeds (almonds, walnuts, pecans, pumpkin seeds, hazelnuts, sunflower seeds, etc.): in a large bowl, dissolve 1 tablespoon sea salt flakes in 700ml filtered water. Rinse and drain 250g nuts, add them to the bowl and cover with a clean cloth. Place in a warm spot for at least 6 hours or up to overnight (softer nuts like cashews, macadamia nuts and pine nuts may require less soaking time). Drain and rinse with fresh water. Use immediately, or store in the fridge in an airtight container and use within one week. If you're storing the nuts or seeds in the fridge, rinse them with fresh water every other day. You can also dehydrate soaked nuts and seeds in a dehydrator or 65°C oven for at least 12 hours, until crispy, then store them in an airtight container at room temperature.

Sprouting is like an extreme version of soaking, in which grains, nuts, seeds or beans are allowed to germinate and eventually sprout, producing roots, a stem and pale, underdeveloped leaves. This boosts the enzyme content, allowing for better absorption of the seeds' powerful nutrients. For example, broccoli sprouts can contain between ten and one hundred times more autophagy-activating sulforaphane than mature broccoli! Start to finish, the sprouting process takes about one week, but will vary depending on what you are sprouting. You'll need a sprouting lid with holes for air circulation (you can buy one online for less than £10 or make your own). Following is a recipe for growing broccoli sprouts, but it can be adapted to anything you are sprouting.

To sprout broccoli seeds, place 2 tablespoons broccoli seeds (be sure to buy broccoli sprouting seeds) in a wide-mouth 1-litre kilner jar. Cover with a few inches of filtered water and cap with a sprouting lid. Set aside in a warm, dark place for at least 8 hours or up to overnight. Drain the liquid and rinse the seeds well. Return the seeds to the jar, cover with the lid and set aside in a warm, dark place. Continue to rinse and drain three or four times a day. You will start to see sprouts in 3 to 4 days. Once the sprouts are about 1 inch long and have yellow leaves, you can move the jar into a sunny spot. Continue to rinse and drain them three or four times a day until the leaves are dark green. The sprouts are now ready to eat. Store in the refrigerator in a plastic bag lined with paper towels for 5 to 7 days.

Fermenting foods may be one of the best ways you can boost autophagy to beef up your glow. The microflora that live in fermented and pre- and probiotic foods create a protective lining in your intestines. They increase antibodies, helping to strengthen your immune system. And they regulate appetite and reduce sugar and refined-carb cravings. Fermenting vegetables is also a great way to preserve them for a longer period of time. This gives you the option to make a large batch of a fermented food and have a ready-to-eat dose of microflora at your disposal. Culinary traditions include fermenting a wide variety of foods, among them grains, to make porridge and breads; milk, to make yogurts and cheeses; and all sorts of vegetables, to make dishes like kimchi and sauerkraut.

To make a basic fermented sauerkraut, shred 1 head cabbage using a food processor fitted with the shredding plate or by hand with a sharp knife. Put the cabbage in a large bowl and add 1 tablespoon sea salt flakes. Use a large wooden spoon to pound the cabbage to release its juices, about

10 minutes. The cabbage should release quite a bit of liquid. Place the cabbage and any liquid in a sterilized wide-mouth kilner jar and press down firmly until the liquid rises above the cabbage. The cabbage needs to be submerged in the liquid for the anaerobic process of fermentation to occur and to prevent the growth of mould. If you need to, add filtered water and place a clean rock on top of the cabbage to act as a weight and keep the cabbage submerged. Cover tightly and set aside at room temperature for 3 days. Transfer to the fridge; eat the sauerkraut immediately or continue to let it ferment in the fridge for up to 1 month.

Enhance Your Flavour: Herbs and Spices

As taste enhancers, herbs and spices can transform foods that might otherwise be considered boring or bland into something delicious and delectable. And even more important, these little additions can have a great influence on autophagy activation. Many culinary herbs and spices contain antioxidant, antimicrobial, anti-inflammatory and anticarcinogenic compounds, boosting your overall health and well-being.

I try to add autophagy-activating herbs and spices to every meal. Here are some of my favourites, which I use in many of the Glow15 recipes starting on page 198.

Cinnamon: It is believed that cinnamon plays a role in lowering blood sugar and LDL cholesterol in people with type-2 diabetes. Try sprinkling it on fruits and veggies – from apples to sweet potatoes – and even on lentils. This powerful spice really perks up side dishes. I also use cinnamon in my youth-boosting AutophaTea (page 199).

Turmeric: You may remember this one from my Powerphenol nutritional supplements. It contains an active component called curcumin that has been shown to reduce inflammation. Try turmeric in soups, stews and curries.

Garlic: Chances are, you have garlic in your kitchen in some form or another. Studies show that just two fresh cloves a week provide anti-cancer benefits, but powdered, minced and granulated garlic also provide excellent flavour. You can make your garlic even more potent by letting it sit for about 10 minutes after chopping, as this allows its active compounds to further develop. Use it in everything from eggs to tuna salad to baked fish.

Oregano: This powerful herb is loaded with antioxidants. It is even believed to help ward off food poisoning, as it can keep harmful bacteria found on food from growing and spreading. Try adding 1½ teaspoons to salad dressings, leafy greens or soups for added flavour.

Rosemary: As you now know, grilling at high temps can produce harmful free radicals that may cause cancer. But a study at Kansas State University found that marinating meat in a rosemary mixture can rev up your resistance. Add 1 tablespoon fresh or dried rosemary to your favourite marinade.

While these are my top five for boosting cellular cleanup, there are countless herbs and spices for you to choose. Just keep in mind the following: fresh is best, and organic is always preferable. I'm not telling you to grow your own – although you can, even just on a windowsill – but do your best to choose fresh, organic herbs in the produce section of your local supermarket. If you do use dried herbs, try mashing them a bit between your fingers to release their essential oils. While dried herbs and spices will keep indefinitely, it's best to restock them at least once a year, as they lose potency over time.

The Proof Is in the Pudding

You now have the tools you need to activate your autophagy, lose weight, rejuvenate your skin and elevate your life. In the next chapter, I lay out a sample fifteen-day plan designed to maximize your results. Get ready to prove to yourself that you deserve to be your first priority. Be bold, and know you can radiate beauty, health, and strength from the inside out.

Glow15 Meals

	HIGH DAYS	LOW DAYS	
Meal 1	AvocaGlow – High Egg15 – High	Intermittent fast *(skip Meal 1)*	
Meal 2	Turkey Lettuce Wrap Salmon Salad Superfood BLT Bowl Loaded Portobello Mushrooms Steak Power Bowl	AvocaGlow – Low Egg15 – Low	
Meal 3	Prawn and Broccoli Stir-Fry Chicken with Satutéed Red Cabbage and Rice Wild Salmon with Mixed Herby Butter Beef Stew Almond Flatbread and Tahini Turkey Veggie Bake Herbed Pork Loin Roast	Mediterranean Chopped Salad Grapefruit Salad Cucumber Salad Veggie Roast Broccoli Soup Pesto Courgetti Powerphenol Purple Smoothie Cauliflower Mushroom Roast Taco Bowl Salad Greens Bowl Honey-Roasted Brussels Sprouts	SIDE DISHES: pick one to pair with your Low-day meal choice Sautéed Broccoli Lemon Asparagus Soup Cauliflower Rice Miso-Glazed Courgettes Spaghetti Squash Veggie Kebab
Snacks	Nutty Chia Clusters Strawberry Coconut Balls Fruit Vegetables	Macadamia Chocolate Bark Asparagus Chips Slow-roasted Cherry Tomatoes Salt-and-Vinegar Kale Crisps Fruit Vegetables	

Chapter 11

Glow15:
The 15-Day Plan

Let's Get Glowing

Whatever your starting weight, number of wrinkles on your skin, work responsibilities or family situation, I know you can glow. You have the power to activate your autophagy to make your body and brain work better and act younger. Glow15 will increase your youth span – the amount of time in years that you live with vibrancy, focus, energy and glowing radiance – helping counter the effects of both the Inevitable and Accelerated Agers (see page 32). In this chapter, I'll outline how to do that for the first fifteen days in a manageable, effective and efficient way. To make it easy, I've detailed exactly what to do and when to do it, so instead of worrying about your schedule, you can focus on how much younger you'll look and feel.

Glow15 Meals

Each day you'll pick one item from meals 1, 2 and 3, and snacks. All you have to do is choose High- or Low-day meals from the list of options that follow. There is plenty of variety and you should never feel deprived. You can have more than one portion of each meal, and you can have more meals than the ones I've outlined below. Your only limitations are on Low days, when you'll be skipping meal 1 in favour of intermittent fasting. After that, feel free to choose any of the low-protein options for meals 2 and 3 and snacks, as you still have eight hours when you can eat. All the recipes (which you'll find starting on page 198) are simple to prepare and quick to make.

The Sample Plan

You're ready now to begin your first fifteen days and get glowing! In these two weeks, by making small changes to how and what you eat, how and when you exercise, how you approach bedtime and what you do to fall asleep, and using the world's most powerful anti-ageing ingredients, you'll see and feel autophagy at work.

This fifteen-day plan is designed to not only kick-start your autophagy, but also work as a blueprint for how you can integrate these lifestyle changes as a regular part of your life. I know you, like so many others, will be motivated and inspired by the youthful results you see and feel.

Here's the full breakdown of what to do and when to do it for the next fifteen days. I've chosen Monday, Wednesday and Friday to be your Low days – this way, your entire weekend can be more flexible. I've also noted that the intermittent fast (IF) on this sample plan begins after your evening meal the night before – ideally around 8 p.m. As long as you have three nonconsecutive Low days every week, both the choice of which days and the time you begin your intermittent fast are up to you. (Again, Low days are the days you fast for 16 hours and restrict your protein.)

And every day – both High and Low – should start with an AutophaTea (page 199). I've put only your first cup in the sample plan, but you can enjoy up to four servings of this youth-boosting drink each day.

Glow15 Meals

Sunday

Morning	AutophaTea
	At least 30 minutes HIIT or RET
	High Meal 1
	Powerphenols
Noon	High Meal 2
	Powerphenols
Evening	High Meal 3
	Powerphenols
	Begin intermittent fasting
	Beauty bath
	Targeted topicals
	Sleep routine for your bird type

TIP

Use social media to control hunger while intermittent fasting. According to a study from Brigham Young University's Marriott School of Management, Instagram or any social media photos of food can boost sensory boredom. In other words, you become tired of eating a food long before you even taste it.

Monday

Day 2   Low Day

Morning	Meditate
	Intermittent fast – skip Meal 1
Noon	Break your intermittent fast
	AutophaTea
	Low Meal 2
	Powerphenols
Evening	Low Meal 3
	Powerphenols
	Targeted topicals
	DIY treatment: Make Cellulite Glow Away
	Sleep routine for your bird type

TIP

Set yourself up for success by meditating first thing today. This is when you need it the most, since your blood pressure and stress hormones are at their highest.

Tuesday

Day 3   High Day

Morning   AutophaTea
               At least 30 minutes HIIT or RET
               High Meal 1
               Powerphenols

Noon   High Meal 2
               Powerphenols

Evening   High Meal 3
               Powerphenols
               Begin intermittent fasting
               Beauty bath
               Targeted topicals
               Sleep routine for your bird type

TIP

Boost the effects of exercise with a hot shower afterward. The heat will boost autophagy and prolong the effects of your workout.

Wednesday

Day 4   Low Day

Morning	Intermittent fast – skip Meal 1
Noon	Break your intermittent fast
	AutophaTea
	Low Meal 2
	Powerphenols
Evening	Low Meal 3
	Powerphenols
	Targeted topicals
	DIY treatment: *Unmask Your Glow*
	Sleep routine for your bird type

TIP

Feeling hangry? Try an extra cup of AutophaTea to help satiate hunger and improve your mood. Remember, you can have up to four cups a day, even on Low days.

Thursday

Day 5 High Day

Morning	AutophaTea
	At least 30 minutes HIIT or RET
	High Meal 1
	Powerphenols
Noon	High Meal 2
	Powerphenols
Evening	High Meal 3
	Powerphenols
	Begin fasting
	Beauty bath
	Targeted topicals
	Sleep routine for your bird type

TIP

Try a high-protein autophagy-boosting snack, like nutty chia clusters (see page 251)

Friday

Day 6   Low Day

Morning	Intermittent fast – skip Meal 1
Noon	Break your intermittent fast
	AutophaTea
	Low Meal 2
	Powerphenols
Evening	Low Meal 3
	Powerphenols
	Targeted topicals
	DIY treatment: Glow-To Facial Scrub
	Sleep routine for your bird type

TIP

Take advantage of your circadian rhythm and your second sleep by taking a nap. Find a dark place and allow yourself 10 minutes of rest to help you recharge and face your day.

Saturday

Day 7 High Day

Morning AutophaTea
     30 minutes HIIT or RET
     High Meal 1
     Powerphenols

Noon   High Meal 2
     Powerphenols

Evening High Meal 3
     Powerphenols
     Beauty bath
     Targeted topicals
     Sleep routine for your bird type

TIP

Craving something sweet? Use 1 teaspoon monk fruit in your AutophaTea to satisfy your hankering.

Sunday

Day 8   High Day

Morning	AutophaTea
	At least 30 minutes HIIT or RET
	High Meal 1
	Powerphenols
Noon	High Meal 2
	Powerphenols
Evening	High Meal 3
	Powerphenols
	Begin intermittent fasting
	Beauty bath
	Targeted topicals
	Sleep routine for your bird type

TIP

Brush your teeth after Meal 3. When your mouth is minty clean and ready for bed, you'll be less likely to raid the fridge and break your intermittent fast.

Monday

Day 9   Low Day

Morning	Intermittent fast – skip Meal 1
Noon	Break your intermittent fast
	AutophaTea
	Low Meal 2
	Powerphenols
Evening	Low Meal 3
	Powerphenols
	Targeted topicals
	DIY treatment: Make Cellulite Glow Away
	Sleep routine for your bird type

TIP

Congrats, you're halfway through your first fifteen days! Celebrate with the Powerphenol resveratrol by having a glass of red wine or a piece of dark chocolate.

Tuesday

Day 10   High Day

Morning   AutophaTea
              At least 30 minutes HIIT or RET
              High Meal 1
              Powerphenols

Noon   High Meal 2
              Powerphenols

Evening   High Meal 3
              Powerphenols
              Begin intermittent fasting
              Beauty bath
              Targeted topicals
              Sleep routine for your bird type

TIP

Boost your beauty bath by adding 2 or 3 drops of bergamot essential oil. Not only is it great aromatherapy, but it's also great for your skin.

Wednesday

Day 11   Low Day

Morning   Intermitent fast – skip Meal 1

Noon   Break your intermittent fast
               AutophaTea
               Low Meal 2
               Powerphenols

Evening   Low Meal 3
               Powerphenols
               Targeted topicals
               DIY treatment: Unmask Your Glow
               Sleep routine for your bird type

TIP

Enjoy some bedroom exercise. Initiate autophagy by having sex tonight. Not only will you reconnect with your partner, but you'll be making each other's cells act younger.

Thursday

Day 12   High Day

Morning	AutophaTea
	At least 30 minutes HIIT or RET
	High Meal 1
	Powerphenols
Noon	High Meal 2
	Powerphenols
Evening	High Meal 3
	Powerphenols
	Begin intermittent fasting
	Beauty bath
	Targeted topicals
	Sleep routine for your bird type

TIP

Try adding one scoop of collagen powder to upgrade your AutophaTea to Beau-Tea! on High Days.

Friday

Day 13   Low Day

Morning	Intermittent fast – skip Meal 1
Noon	Break your intermittent fast
	AutophaTea
	Low Meal 2
	Powerphenols
Evening	Low Meal 3
	Powerphenols
	Targeted topicals
	DIY treatment: Glow-To Facial Scrub
	Sleep routine for your bird type

TIP

Fri-yay! You made it to the weekend. Have some dark chocolate (at least 70% cocoa) to celebrate and activate your autophagy.

Saturday

Day 14 High Day

Morning	AutophaTea
	At least 30 minutes HIIT or RET
	High Meal 1
	Powerphenols
Noon	High Meal 2
	Powerphenols
Evening	High Meal 3
	Powerphenols
	Beauty bath
	Targeted topicals
	Sleep routine for your bird type

TIP

If you haven't already done so, push yourself to exercise longer (up to 80 minutes), then reward yourself with a massage. Either schedule one professionally or have your partner give you one. Not only will you boost your autophagy, you'll be more relaxed the rest of the weekend.

Sunday

Day 15   High Day

Morning   AutophaTea
30 minutes HIIT or RET
High Meal 1
Powerphenols

Noon   High Meal 2
Powerphenols

Evening   High Meal 3
Powerphenols
Begin intermittent fasting
Beauty bath
Targeted topicals
Sleep routine for your bird type

TIP
Get out your worksheet – it's day 15. Time to track your success!

Admire your new glow!

Go naked – OK, makeup free!

Try on clothes in a smaller size.

Keep on glowing – plan your next fifteen days!

The After-Glow

You've made it through the first fifteen days, and there's a good chance you've tried new foods, nutritional supplements, beauty treatments, exercises and sleep techniques. Some of those changes may have been a little uncomfortable at first, but now that you're familiar with all the facets of the programme, I hope it's easier and more enjoyable. There is also a good chance that you experienced some wonderful benefits already – from weight loss to smoother skin to better mental clarity and more energy. You've started to activate your autophagy – that's amazing, and it doesn't have to stop. Glow15 was designed for you to see initial changes in fifteen days, but its principles are intended to be practiced as a lifestyle.

As you continue, feel free to repeat this fifteen-day cycle as is, or with adjustments as needed. Just keep in mind the Glow15 guidelines, and from there, have fun experimenting. Here are some examples:

- **Experiment with recipes.** Modify the recipes in this book to your liking, using autophagy-boosting ingredients. Try new foods and combinations, using the recipes starting on page 198 to help guide you.

- **Experiment with your own IFPC cycles.** Now that you know how three Low days work for you, enjoy the freedom to find out if you respond better to more or fewer days. Maybe you'll want to try four Low days, or maybe your body actually prefers two. Also, you can tinker with the times of your intermittent fast: try starting earlier the night before if you get hungry in the morning, or try the opposite if you like eating later in the evening, pushing your intermittent fast so it ends later the next day. If you choose to experiment, it's best to do so in the same fifteen-day increments so you can best assess your results.

- **Experiment with new exercises.** Now that you understand the principles of HIIT and RET, you can be a little creative with your workouts. That may mean trying different classes or making up your own routines – the point is to challenge yourself. Keep in mind that as you become stronger and more flexible, you can try more advanced exercises and increase the time you spend exercising – remember, while 30 minutes will activate autophagy, you can boost your results with up to 80 minutes of exercise. No need to go past that – just doing these workouts already makes you Superwoman!

Frequently Asked Questions

What should I do when I feel hungry when I'm intermittent fasting on a Low day?

As you start Glow15, it's common to feel hungry during intermittent fasting on Low days. I've found the following to be helpful:

- Keep hydrated! It's perfectly fine to have water, tea or coffee with nothing added during your intermittent fast.
- Make sure you are eating enough on High days. This may not sound like it would help you on a Low day, but it is actually a benefit of IFPC. Excessive hunger may be a sign that you are not eating enough overall or not enough of the nutrient-dense foods necessary to balance the hormones involved in hunger. During High days, it is important to consume enough protein to ensure your body is getting what it needs. Also, don't skimp on fibre-packed vegetables or satiating fat at mealtimes, as these foods are key to success.
- Keep in mind that the more "toxic" or insulin resistant your body is at the beginning of your Glow15 programme, the harder it may be to adjust. But as your system secretes less insulin, your blood sugar will steady and naturally regulate your hunger.

Do I eat fewer calories on a Low day compared to a High day?

No – and on Glow15, you do not have to, nor should you, count calories. Glow15 is about the quality of the nutrients you are consuming. This programme doesn't consider the calorie to be an accurate predictor of success. While caloric restriction is proven to activate autophagy, I factored that in when creating IFPC. Your 16-hour intermittent fast will boost cellular cleansing, and when you're fasting, you're certainly restricting calories. And, as you know, the key to optimizing autophagy is turning it on and off, so after your intermittent fast there is no reason to continue to withhold nutrients.

I heard not eating breaks down my muscles. Should I be worried about intermittent fasting?

As long as you are maintaining the rhythms of IFPC suggested in this programme, you will actually gain muscle, not lose it. The fasting state is a catalyst for the physiological mechanism that helps your body convert protein into muscle. When you are fasting, you are actually setting the stage for muscle regeneration and growth!

Does intermittent fasting cause my body to store fat?

No. Because the intermittent fast is relatively short (16 hours is a short block of time in comparison to other fasting practices), the body's alarm system will not sound or call in the hormonal troops to preserve energy via fat. A quick and short-lived spike in stress hormones such as cortisol and adrenaline may actually help to promote fat burning during a short fast such as IFPC, and will not promote fat storage. Fat storage occurs when elevated insulin is prolonged, cortisol is consistently triggered, and the two are simultaneously elevated. IFPC actually cools the very inflammation upon which fat storage and stubborn weight loss are dependent.

If I eat fat, will I get fat?

No. On the contrary, it's the quality and quantity of carbohydrates that determine how much fat we accumulate. Carbohydrates drive insulin production, and insulin drives fat accumulation. Eating fat and using fat for fuel helps you burn more calories and speeds up your metabolism.

Isn't it right to eat three square meals a day?

The "three meals a day" ritual is culturally rooted – not biologically. It started during the Industrial Revolution when all classes of people, from the poor to the wealthy, began eating a meal before going to work, a midday feast and then a post-workday meal to celebrate the day's labour. This pattern of eating became the norm. Then, in the late nineteenth century, when John Kellogg invented breakfast cereal, the food business capitalized on breakfast being the "most important meal of the day". But the most important meal of the day shouldn't be arbitrary or dictated by big business. On this programme, your most important meal of the day will be defined by the nutrients and polyphenols you use to upgrade your biology.

If I eat right, I don't need to take supplements, right?

While it seems like eating a clean, nutrient-dense diet would negate the need for supplements, modern agricultural practices and poor soil leave our food with less available minerals, vitamins and phytonutrients, and modern living means higher stress levels. Taking some key high-quality supplements is a major factor to augmenting your success. It's not that you won't get success if you *don't* take them, but again, you can think of supplements as a nutritional insurance policy, for boosted success and faster results if you do.

How important is eating organic, free-range, grass-fed animal protein, and will it impact my results if I don't?

As best as your budget allows, choose the least toxic food to reap the biggest payoffs from the Glow15 programme. Food that is as close to its original state in nature as possible, without toxic chemicals, synthetics, food dyes (you read that right – food dye is used in fish to make farmed salmon look wild – yucky, right?), pesticides, herbicides, fungicides, antibiotics and hormones, will be much easier for your body to assimilate. Choosing wholesome organic and grass-fed, pasture-raised or free-range animal protein allows you to avoid inflammation.

Why didn't I feel the glow after fifteen days?

While countless women have given "glowing" testimonials of the programme, attesting to how it helped improve and enhance their lives, and despite being scientifically proven to help you lose weight, look younger, sleep better and feel more focused and energized, Glow15 is not a magic bullet.

But I believe you should get the results you want, and that's why I'm constantly adjusting and upgrading the plan. In search of why there was not 100 per cent satisfaction with the plan, I've asked the questions, looked at the data and listened to women from the clinical study, my inner circle of guinea pigs and the growing number of women who have tried it. And I've found some compelling reasons for why not everyone feels the glow after fifteen days – along with solutions to get you glowing.

You could have done better. Please understand, I do not mean to imply in any way that you did not try your best or follow the plan. This is not your fault, and it does not make you a failure. But you may not have followed the full intent of the programme. Perhaps you made an adjustment – be it to IFPC, exercise, your sleep or your beauty routine – because you needed to, wanted to or thought it wouldn't make a difference. Unfortunately, these modifications in the first fifteen days really do influence your results, as do a lack of planning or preparation, mediocre commitment and inadequate effort. Only you know if you were not truly dedicated to Glow15, and if you could have made a bigger effort. But I want you to know *you can still do it*. I believe if you try again, you will succeed!

Fifteen days wasn't long enough. While the majority of women who've tried Glow15 have seen significant results in just two weeks, the fact is that for some of us, it may take more time. You may remember Lindsay from chapter 1, a participant in the Jacksonville University study (see page 23). While she noticed immediate changes in her skin, her number on the scales barely moved during the first fifteen days of the programme. But by day 60, Lindsay had dropped four dress sizes. Again, fifteen days is how long it will take to see initial changes, but Glow15 is a lifestyle, and committing to it means seeing changes for life. This programme is designed to be measured in increments – day 0, day 15, day 30, etc. If your worksheet doesn't reflect all the changes you hoped for in the first fifteen days, know that it can – it *will* – still happen. (But also make sure you are measuring the best markers . . .)

You're not measuring correctly. Maybe you really wanted to lose inches – but you didn't get to pull your tape measure tighter, or as tight as you wanted, so you assume the programme failed. Did you take into account what else has happened since you started Glow15? Have you paid attention to whether you're falling asleep easier or staying asleep longer? Has your energy improved and become more consistent? Are you stronger? Do you have more stamina? Does your skin look clearer and maybe even glow? While we all have our specific areas we hope to change, don't lose sight of the other results you have achieved or are still achieving. Focus on some of the unexpected changes – the big and small ones – and you may be pleasantly surprised to find all the ways you truly are glowing.

That said, I get it – I can't blame you for being disappointed you haven't seen the results you were expecting to see. Especially when all you have to do is turn the pages of this book to find examples of women just like you who got what you wanted. Well, you don't need me to tell you that you are special. You know that your DNA, your hormone levels, your family dynamic, your work responsibilities and your anxieties can all impact your experience on Glow15. Bottom line: you can have similar traits to another person, but in the end you are unique and your results will be exclusively yours.

All this is to say, if you're not satisfied with your results on Glow15, I urge you to go back and give it another go (to glow!). Perhaps you'll find different benefits or improved results or a new motivation to get glowing.

Chapter 12

Glow15 Recipes

My favourite Glow15 recipes are easy to make and, more importantly, delicious — because I believe in not only loving your food, but also loving to make your food. Each recipe is designed to follow the Glow15 plan, helping you activate your autophagy. The foods needed are all at your local supermarket. The preparation is either minimal or can be done in advance. And the practicality of being able to freeze for future use was considered. To create these recipes, I consulted with nutritionists, chefs and home cooks just like you. With their help, I've also included the macronutrient breakdown (fat, protein and carbohydrate) for each dish. I purposely left out the calorie count for each recipe because on Glow15 we don't count calories. But we do count taste, so enjoy!

Key for Autophagy Activators in Glow15

Polyphenols	PO	Saponins	SA
Sphingolipids	SP	Vitamin C, D, E, K	VIT
Omega-3	O3	Spermidine	SM
Sulforaphane	SU	Probiotics	PB

Use this key to learn exactly which autophagy activators are at play in the recipes in this chapter.

AutophaTea

Autophagy activators: SP, VIT, PO

Makes 1 serving • Prep time: 5 minutes

Your ultimate youth boost. Enjoy up to 1 litre of this delicious and nutritious autophagy-activating drink daily. On both High and Low days, it will keep you satiated, focused and energized. Check the ingredients of your Earl Grey tea: make sure it is made with real bergamot oil, not artifical bergamot flavouring.

Note: Start with 1 teaspoon coconut oil per 250ml, and work your way up to 1 to 2 tablespoons over several days.

1 green tea bag
1 whole citrus bergamot Earl Grey tea bag
1 cinnamon stick
1 tablespoon raw coconut oil (see Note)
1 teaspoon monk fruit (optional)

1. In a kettle or small saucepan, bring 250–350ml water to the boil. Pour the water into a large mug and add the tea bags and cinnamon stick. Let infuse for at least 3 minutes (the longer the better), then remove and discard the tea bags.
2. Add the coconut oil and stir it in using the cinnamon stick.
3. Mix it all together for 20–30 seconds. You can also blend the tea to help mix the flavours and emulsify the oil.
4. If you like, sweeten the tea with the monk fruit.

Nutritional analysis per serving (1 tea): fat 14g, protein 0g, carbohydrate 0g
Nutritional analysis per serving with monk fruit (1 tea): fat 14g, protein 0g, carbohydrate 4g

AvocaGlow

Autophagy activators: SP, VIT, PO

Makes 1 serving • Prep time: 5 minutes

This is the Glow15 go-to meal to get fat first with nutrient-dense autophagy activators. It's easy to make and can be enjoyed on both High and Low days. My children request it for their breakfast and I love it to break my fast, too.

2 teaspoons avocado oil
1 teaspoon fresh lemon juice
Sea salt flakes
Chilli flakes
½ avocado

AvocaGlow High Option 1

2 medium eggs, cooked to your liking
25g feta cheese, crumbled

AvocaGlow High Option 2

115g smoked salmon
25g goats' cheese, crumbled

1. In a small bowl, whisk together the avocado oil, lemon juice, salt and chilli flakes.
2. Drizzle the mixture over the avocado and enjoy. On High days, add your favourite protein option.

Nutritional analysis per serving (½ avocado): fat 20g, protein 1g, carbohydrate 6g, net carbs 1g

Nutritional analysis per serving, option 1 (½ avocado): fat 34g, protein 19g, carbohydrate 8g

Nutritional analysis per serving, option 2 (½ avocado): fat 33g, protein 24g, carbohydrate 7g

Egg15 Recipes

One of my favourite parts of Glow15 is the introduction of Egg15. These fifteen recipes created for both High and Low days are easy to prepare, can be made well in advance, are convenient for travel and storage – and most importantly, they're delicious!

You prepare and bake these eggs in muffin tins, which gives you a perfect portion size every time (and I like the shape!). And they're designed for you to be able to prepare both High- and Low-day versions of Egg15 at once – this not only maximizes your time, but also the use of eggs. You're able to do so because egg whites are high in protein, so you will only use egg yolks on Low days, but you can use those extra whites for High days (if you make six servings of an Egg15 Low recipe, you will have six egg whites left over to use in your Egg15 High recipe).

Make extra and store them in the freezer so you won't have to make any the following week. To freeze the egg muffins, leave to them cool fully at room temperature or in the fridge, place them in a freezer-safe plastic food bag and freeze for up to three months. To serve them from the freezer, allow them to thaw overnight in the fridge and then warm in the microwave on a plate or in a glass bowl for 20–30 seconds. (Hint: adding a few drops of water to the eggs or even on the plate before reheating helps hydrate the egg muffins and gives them a 'fresh' texture.)

The best part: each Egg15 serving is two muffins.

Pizza Omelette Egg Muffins (High)

Autophagy activators: SP, SU, SM, PO, VIT, SA

Makes 3 servings (6 muffins)
Prep time: 10 minutes • Cook time: 20 minutes

1½ teaspoons avocado oil
6 medium eggs
6 egg whites (reserved from Egg15 Low recipes)
Pinch of sea salt flakes
Pinch of freshly ground black pepper
1 medium tomato, diced
20g fresh basil, chopped
1 tablespoon chopped fresh oregano, or ½ teaspoon dried
55g fresh mozzarella cheese, grated

1. Preheat the oven to 180°C/Gas Mark 4. Grease six holes of a muffin tin with the oil.
2. In a large bowl, whisk together the eggs, egg whites, salt and pepper. Set aside.
3. Evenly distribute the tomato, basil and oregano between the greased muffin holes.
4. Scatter 25g of the cheese over the tomatoes and herbs.
5. Add 75ml of the egg mixture to each muffin hole, taking care not to overfill them.
6. Scatter the remaining cheese over the egg, dividing it evenly.
7. Bake for 15–20 minutes until the eggs are set.
8. Serve immediately, or leave to cool completely and freeze (see page 201).

Nutritional analysis per serving (2 egg muffins, not including extra egg whites): fat 17g, protein 17g, carbohydrate 4g
Nutritional analysis per serving (2 egg muffins, including extra egg whites): fat 17g, protein 25g, carbohydrate 4g

Broccoli Cheddar Egg Muffins (High)

Autophagy activators: SP, SU, SM, VIT, PO

Makes 3 servings • Prep time: 10 minutes • Cook time: 20 minutes

1½ teaspoons avocado oil
6 medium eggs
6 egg whites (reserved from Egg15 Low recipes)
Pinch of sea salt flakes
Pinch of freshly ground black pepper
1 tablespoon chopped fresh chives (optional)
70g small broccoli florets
55g Cheddar cheese, grated

1. Preheat the oven to 180°C/Gas Mark 4. Grease six holes of a muffin tin with the oil.
2. In a large bowl, whisk together the eggs, egg whites, salt, pepper and chives (if using). Set aside.
3. In a medium-sized bowl, toss together the broccoli and cheese.
4. Divide the broccoli-cheddar mixture evenly between the greased muffin holes.
5. Add 75ml of the egg mixture to each muffin hole, taking care not to overfill them.
6. Bake for 15–20 minutes until the eggs are set.
7. Serve immediately, or let cool completely and freeze (see page 201).

Nutritional analysis per serving (2 egg muffins, not including extra egg whites): fat 18g, protein 17g, carbohydrate 4g, net carbs 3g
Nutritional analysis per serving (2 egg muffins, including extra egg whites): fat 18g, protein 24g, carbohydrate 4g, net carbs 3g

Salmon, Onion, Cream Cheese and Dill Egg Muffins (High)

Autophagy activators: SP, O3, SM, VIT, PO

Makes 3 servings • Prep time: 10 minutes • Cook time: 20 minutes

2 teaspoons avocado oil
6 medium eggs
6 egg whites (reserved from Egg15 Low recipes)
Pinch of sea salt flakes
Pinch of freshly ground black pepper
½ onion, chopped
115g thinly sliced smoked salmon, cut into 1-cm-wide strips
115g goats' cheese, cut into 1-cm pieces
1 tablespoon chopped fresh dill

1. Preheat the oven to 180°C/Gas Mark 4. Grease six holes of a muffin tin with 1½ teaspoons of the oil.
2. In a large bowl, whisk together the eggs, egg whites, salt and pepper. Set aside.
3. In a medium-sized frying pan, heat ½ teaspoon oil over a medium-high heat. Add the onion and cook, stirring, until translucent, 5–8 minutes. Leave to cool.
4. Layer the onion, salmon and cheese in the muffin holes.
5. Add 75ml of the egg mixture to each muffin hole, taking care not to overfill them.
6. Scatter the dill evenly over the egg.
7. Bake for 15–20 minutes until the eggs are set.
8. Serve immediately, or cool completely and freeze (see page 201).

Nutritional analysis per serving (2 egg muffins, not including extra egg whites): fat 21g, protein 27g, carbohydrate 3g

Nutritional analysis per serving (2 egg muffins, including extra egg whites): fat 17g, protein 35g, carbohydrate 4g

Spinach, Bacon and Goats' Cheese Egg Muffins (High)

Autophagy activators: SP, PO, VIT, SM

Makes 3 servings • Prep time: 15 minutes • Cook time: 20 minutes

2 teaspoons avocado oil
2 rashers nitrite-free bacon
25g fresh spinach, chopped
6 medium eggs
6 egg whites (reserved from Egg15 Low recipes)
Pinch of sea salt flakes
Pinch of freshly ground black pepper
2 tablespoons crumbled goats' cheese
2 tablespoons chopped fresh parsley

1. Preheat the oven to 180°C/Gas Mark 4. Grease six moles of a muffin tin with 1½ teaspoons of the oil.
2. In a medium-sized frying pan, cook the bacon over a medium heat until golden. Set aside on kitchen towel to drain, then roughly chop.
3. In the same pan, heat ½ teaspoon oil over a medium-high heat. Add the spinach and cook until wilted, 1–2 minutes. Set aside.
4. In a bowl, whisk together the eggs, egg whites, salt and pepper.
5. Divide the spinach between the muffin holes, then the bacon.
6. Add 75ml of the egg mixture to each muffin hole, taking care not to overfill them.
7. Scatter the goats' cheese and parsley evenly over the egg.
8. Bake for 15–20 minutes until the eggs are set.
9. Serve immediately, or leave to cool completely and freeze (see page 201).

Nutritional analysis per serving (2 egg muffins, not including extra egg whites): fat 14g, protein 15g, carbohydrate 1g
Nutritional analysis per serving (2 egg muffins, including extra egg whites): fat 17g, protein 23g, carbohydrate 4g

Greek Egg Muffins (High)

Autophagy activators: SP, PO, SA, SM, VIT

Makes 3 servings • Prep time: 10 minutes • Cook time: 20 minutes

2 teaspoons avocado oil
55g spinach, chopped
6 medium eggs
6 egg whites (reserved from Egg15 Low recipes)
Pinch of sea salt flakes
Pinch of freshly ground black pepper
6 pitted black olives, chopped
1 medium tomato, chopped
1½ teaspoons dried oregano
70g feta cheese, crumbled

1. Preheat the oven to 180°C/Gas Mark 4. Grease six holes of a muffin tin with 1½ teaspoons of the oil.
2. In a medium-sized frying pan, heat ½ teaspoon oil over a medium-high heat. Add the spinach and cook until wilted, 1–2 minutes. Set aside to cool.
3. In a bowl, whisk the eggs, egg whites, salt and pepper together.
4. Layer the spinach, olives and tomato in the muffin holes.
5. Add 75ml of the egg mixture to each muffin hole, taking care not to overfill them.
6. Scatter the oregano and feta cheese evenly over the egg.
7. Bake for 15–20 minutes until the eggs are set.
8. Serve immediately, or leave to cool completely and freeze (see page 201).

Nutritional analysis per serving (2 egg muffins, not including extra egg whites): fat 22g, protein 21g, carbohydrate 6g, net carbs 5g
Nutritional analysis per serving (2 egg muffins, including extra egg whites): fat 22g, protein 29g, carbohydrate 6g, net carbs 5g

BLT Egg Muffins (High)

Autophagy activators: SP, PO, VIT, SA, SM, PO

Makes 3 servings • Prep time: 15 minutes • Cook time: 20 minutes

1½ teaspoons avocado oil
2 rashers nitrite-free bacon
6 medium eggs
6 egg whites (reserved from Egg15 Low recipes)
25g rocket, roughly chopped
1 medium tomato, chopped
Pinch of sea salt flakes
Pinch of freshly ground black pepper
2 tablespoons chopped fresh parsley

1. Preheat the oven to 180°C/Gas Mark 4. Grease six holes of a muffin tin with the oil.
2. In a medium-sized frying pan, cook the bacon over a medium-high heat until golden. Set aside on kitchen paper to drain and cool. Roughly chop the bacon and set aside.
3. In a large bowl, whisk the eggs, egg whites, rocket, tomato, salt and pepper together.
4. Add 75ml of the egg mixture to each muffin hole, taking care not to overfill them.
5. Scatter the bacon and parsley evenly over the egg.
6. Bake for 15–20 minutes until the eggs are set.
7. Serve immediately, or leave to cool completely and freeze (see page 201).

Nutritional analysis per serving (2 egg muffins, not including extra egg whites): fat 14g, protein 15g, carbohydrate 4g, net carbs 3g
Nutritional analysis per serving (2 egg muffins, including extra egg whites): fat 17g, protein 23g, carbohydrate 4g, net carbs 3g

Sausage, Kale and Onion Egg Muffins (High)

Autophagy activators: SP, PO, VIT, SU, SM

Makes 3 servings • Prep time: 15 minutes • Cook time: 20 minutes

2 teaspoons avocado oil
115g sausage meat
½ onion, chopped
100g kale leaves, very thinly sliced
Sea salt and freshly ground black pepper
6 medium eggs
6 egg whites (reserved from Egg15 Low recipes)

1. Preheat the oven to 180°C/Gas Mark 4. Grease six holes of a muffin tin with 1½ teaspoons of the oil.
2. In a medium-sized frying pan, heat ½ teaspoon oil over a medium-high heat. Add the sausage and cook, breaking up the meat with a wooden spoon, until evenly browned. Set aside.
3. In the same pan, cook the onion over a medium heat until translucent, 5–8 minutes. Reduce the heat to low, add the kale and gently cook until it wilts. Leave to cool. Season with salt and pepper.
4. In a large bowl, whisk the eggs, egg whites and salt and pepper together.
5. Divide the sausage evenly between the greased muffin holes.
6. Divide the onion-kale mixture evenly over the sausage.
7. Add 75ml of the egg mixture to each muffin hole, taking care not to overfill them.
8. Bake for 15–20 minutes, until the eggs are set.
9. Serve immediately, or cool completely and freeze (see page 201).

Nutritional analysis per serving (2 egg muffins, not including extra egg whites): fat 23g, protein 19g, carbohydrate 4g

Nutritional analysis per serving (2 egg muffins, including extra egg whites): fat 17g, protein 27g, carbohydrate 4g

Bacon, Spinach and Cheese Egg Muffins (High)

Autophagy activators: SP, SM, PO, VIT

Makes 3 servings • Prep time: 15 minutes • Cook time: 20 minutes

1½ teaspoons avocado oil
2 rashers nitrite-free bacon
55g spinach, chopped
6 medium eggs
6 egg whites (reserved from Egg15 Low recipes)
Pinch of sea salt flakes
Pinch of freshly ground black pepper
55g Cheddar cheese, grated

1. Preheat the oven to 180°C/Gas Mark 4. Grease six holes of a muffin tin with the oil.
2. In a medium-sized frying pan, cook the bacon over a medium heat until golden. Drain on kitchen paper and leave to cool. Roughly chop the bacon and set aside.
3. In the bacon drippings, cook the spinach over a medium heat, stirring, until wilted.
4. In a medium-sized bowl, whisk the eggs, egg whites, salt and pepper together.
5. Add the bacon, spinach and cheese to the eggs and stir to combine.
6. Add 75ml of the egg mixture to each muffin hole, taking care not to overfill them.
7. Bake for 15–20 minutes until the eggs are set.
8. Serve immediately, or leave to cool completely and freeze (see page 201).

Nutritional analysis per serving (2 egg muffins, not including extra egg whites): fat 19g, protein 18g, carbohydrate 2g
Nutritional analysis per serving (2 egg muffins, including extra egg whites): fat 17g, protein 26g, carbohydrate 4g

Meat Lover's Egg Muffins (High)

Autophagy activators: SP, SM, VIT, PO

Makes 3 servings • Prep time: 15 minutes • Cook time: 20 minutes

2½ teaspoons avocado oil
115g beef mince
Sea salt flakes and freshly ground black pepper
6 medium eggs
6 egg whites (reserved from Egg15 Low recipes)
55g sliced ham, chopped
55g mature Cheddar cheese, grated
25g chopped fresh parsley

1. Preheat the oven to 180°C/Gas Mark 4. Grease six holes of a muffin tin with 1½ teaspoons of the oil.
2. In a medium-sized frying pan, heat 1 teaspoon oil. Add the beef mince and cook, breaking up the meat with a wooden spoon as it cooks, until the beef is browned. Season with salt and pepper and set aside.
3. In a large bowl, whisk the eggs, egg whites and salt and pepper together.
4. Layer the beef, ham, cheese and parsley evenly between the greased muffin holes.
5. Add 75ml of the egg mixture to each muffin hole, taking care not to overfill them.
6. Bake for 15–20 minutes until the eggs are set.
7. Serve immediately, or leave to cool completely and freeze (see page 201).

Nutritional analysis per serving (2 egg muffins, not including extra egg whites): fat 26g, protein 28g, carbohydrate 3g
Nutritional analysis per serving (2 egg muffins, including extra egg whites): fat 17g, protein 36g, carbohydrate 4g

Chard, Tomato and Onion Egg Muffins (Low)

Autophagy activators: PO, VIT, SP, SA

Makes 3 servings • Prep time: 10 minutes • Cook time: 20 minutes

2 teaspoons avocado oil
½ onion, chopped
70g chard, roughly chopped
6 egg yolks (reserve the whites for Egg15 High recipes)
Sea salt flakes and freshly ground black pepper
6 cherry tomatoes, halved

1. Preheat the oven to 180°C/Gas Mark 4. Grease six holes of a muffin tin with 1½ teaspoons of the oil.
2. In a large frying pan, heat ½ teaspoon oil over a medium-high heat. Add the onion and cook, stirring, until translucent, 5–8 minutes. Turn down the heat to low, add the chard and cook, stirring, until the chard wilts. Set aside to cool.
3. In a medium-sized bowl, gently whisk the egg yolks to combine.
4. Add the onion-chard mixture to the egg yolks and season with salt and pepper. Stir well.
5. Divide the egg mixture evenly between the greased muffin holes, taking care not to overfill them. Top each cup with 2 tomato halves.
6. Bake for 15–20 minutes until the eggs are set.
7. Serve immediately, or leave to cool completely and freeze (see page 201).

Nutritional analysis per serving (2 egg muffins): fat 10g, protein 6g, carbohydrate 3g,
net carbs 2g

Mushroom and Onion Egg Muffins (Low)

Autophagy activators: SP, SM, PO

Makes 3 servings • Prep time: 10 minutes • Cook time: 20 minutes

2 teaspoons avocado oil
½ onion, chopped
70g button mushrooms, sliced
Pinch of sea salt flakes
Pinch of freshly ground black pepper
6 egg yolks
2 tablespoons chopped fresh parsley

1. Preheat the oven to 180°C/Gas Mark 4. Grease six holes of a muffin tin with 1½ teaspoons of the oil.
2. In a medium-sized frying pan, heat ½ teaspoon oil over a medium-high heat. Add the onion and cook, stirring, until translucent, 5–8 minutes. Transfer the onions to a bowl and set aside.
3. In the same pan, cook the mushrooms over a medium heat until golden brown, about 10 minutes.
4. Drain the mushrooms and add them to the bowl with the onion. Season with salt and pepper. Mix well.
5. In a separate bowl, gently whisk the egg yolks.
6. Add the egg yolks to the onion and mushrooms. Mix well.
7. Divide the egg mixture evenly between the greased muffin holes, taking care not to overfill them.
8. Scatter the egg evenly with the parsley.
9. Bake for 15–20 minutes, until the eggs are set.
10. Serve immediately, or cool completely and freeze (see page 201).

Nutritional analysis per serving (2 egg muffins): fat 10g, protein 6g, carbohydrate 3g,
net carbs 2g

Olive, Pepper and Onion Egg Muffins (Low)

Autophagy activators: SP, SA, PO, VIT

Makes 3 servings • Prep time: 10 minutes • Cook time: 20 minutes

2 teaspoons avocado oil
½ onion, chopped
1 pepper, finely diced
6 black olives, pitted and chopped
Pinch of sea salt flakes
Pinch of freshly ground black pepper
6 egg yolks

1. Preheat the oven to 180°C/Gas Mark 4. Grease six holes of a muffin tin with 1½ teaspoons of the oil.
2. In a medium-sized frying pan, heat ½ teaspoon oil over a medium-high heat. Add the onion and cook, stirring, until translucent, 5–8 minutes. Transfer the onion to a bowl and set aside.
3. In the same pan, cook the diced pepper over medium heat until soft, 6–8 minutes. Add the diced pepper to the bowl with the onion, add the olives, and season with salt and black pepper. Mix well.
4. In a separate bowl, gently whisk the egg yolks.
5. Add the egg yolks to the bowl with the vegetables and mix well.
6. Divide the egg mixture evenly between the greased muffin holes, taking care not to overfill them.
7. Bake for 15–20 minutes until the eggs are set.
8. Serve immediately, or leave to cool completely and freeze (see page 201).

Nutritional analysis per serving (2 egg muffins): fat 11g, protein 6g, carbohydrate 5g, net carbs 4g

Kale and Jalapeño Egg Muffins (Low)

Autophagy activators: SP, PO, VIT, SA, SU

Makes 3 servings • Prep time: 10 minutes • Cook time: 20 minutes

1½ teaspoons avocado oil
100g kale, stalks removed and finely chopped
1 fresh jalapeño, seeds removed and finely diced
½ teaspoon ground cumin
Pinch of sea salt flakes
Pinch of freshly ground black pepper
6 egg yolks
2 tablespoons chopped fresh coriander

1. Preheat the oven to 180°C/Gas Mark 4. Grease six holes of a muffin tin with the oil.
2. Cook the kale in a steam basket until lightly steamed, 2–3 minutes. Remove from the heat and place in a mixing bowl.
3. Add the jalapeño, cumin, salt and pepper to the bowl with the kale. Stir to combine. Set aside.
4. In a separate bowl, gently whisk together the egg yolks.
5. Dived the kale-jalapeño mixture evenly between the greased muffin holes. Scatter coriander evenly over the kale mixture.
6. Add 75ml of the egg mixture to each muffin hole, taking care not to overfill them.
7. Bake for 15–20 minutes until the eggs are set.
8. Serve immediately, or leave to cool completely and freeze (see page 201).

Nutritional analysis per serving (2 egg muffins): fat 10g, protein 5g, carbohydrate 2g, net carbs 1g

Herb and Asparagus Egg Muffins (Low)

Autophagy activators: SP, PO, VIT

Makes 3 servings • Prep time: 10 minutes • Cook time: 20 minutes

2 teaspoons avocado oil
60g asparagus, cut into 5-mm pieces
1 teaspoon sea salt flakes
6 egg yolks
Pinch of freshly ground black pepper
2 teaspoons chopped fresh chives
2 teaspoons chopped fresh parsley

1. Preheat the oven to 180°C/Gas Mark 4. Grease six holes of a muffin tin with 1½ teaspoons of the oil.
2. In a medium-sized frying pan, heat ½ teaspoon oil over a medium heat. Add the asparagus and ½ teaspoon of the salt and cook, stirring, until softened, 5–10 minutes.
3. While the asparagus is cooking, in a medium-sized bowl, gently whisk the egg yolks and remaining ½ teaspoon salt and pepper together.
4. Divide the asparagus evenly between the greased muffin holes. Scatter evenly with the chives and parsley.
5. Add 75ml of the egg mixture to each muffin hole, taking care not to overfill them.
6. Bake for 15–20 minutes until the eggs are set.
7. Serve immediately, or leave to cool completely and freeze (see page 201).

Nutritional analysis per serving (2 egg muffins): fat 10g, protein 5g, carbohydrate 2g,
net carbs 1g

Curried Pepper and Onion Egg Muffins (Low)

Autophagy activators: SP, SA, PO, VIT

Makes 3 servings • Prep time: 10 minutes • Cook time: 20 minutes

2 teaspoons avocado oil
1 red pepper, chopped
½ onion, chopped
6 egg yolks
Pinch of sea salt flakes
Pinch of freshly ground black pepper
¼ teaspoon curry powder

1. Preheat the oven to 180°C/Gas Mark 4. Grease six holes of a muffin tin with 1½ teaspoons of the oil.
2. In a large frying pan, heat ½ teaspoon oil over a medium-high heat. Add the redd pepper and onion and cook, stirring, until the onion is translucent, 5–8 minutes. Set aside to cool.
3. In a medium-sized bowl, gently whisk the egg yolks.
4. Add the vegetables to the egg yolks, then stir in the salt, black pepper and curry powder.
5. Divide the egg mixture evenly betewen the greased muffin holes, taking care not to overfill them.
6. Bake for 15–20 minutes until the eggs are set.
7. Serve immediately, or leave to cool completely and freeze (see page 201).

Nutritional analysis per serving (2 egg muffins): fat 10g, protein 6g,
carbohydrate 6g,
net carbs 1g

High-Day Recipes

Turkey Lettuce Wrap

Autophagy activators: SA, PO, SP, VIT

Makes 1 serving • Prep time: 5 minutes

Perfect for when you need a quick lunch on high-protein days. Swapping in fresh, nutrient-dense lettuce makes a wonderful substitute for a processed and refined wrap. Spring greens and cabbage leaves are great options, too – steam them first, or try them raw! Serve with a plate of freshly cut vegetables and dip from the AutophaSauces section (pages 254–60).

1 large cos lettuce leaf
1–2 tablespoons avocado mayonnaise, tahini or guacamole
115g roast turkey breast slices
2 rashers nitrite-free bacon, cooked
¼ avocado, sliced
30g salad leaves or rocket
1 small tomato, sliced
Chopped fresh basil (optional)

1. Place the cos leaf on a plate.
2. Spread the avocado mayonnaise, tahini or guacamole on the leaf. Top with the turkey, bacon, avocado, salad leaves and tomato.
3. Scattered with basil, if liked. Wrap the sides of the cos leaf around the filling and enjoy immediately.

Nutritional analysis per serving (1 wrap, including 2 tablespoons avocado-based mayo): fat 32g, protein 27g, carbohydrate 7g

Salmon Salad

Autophagy activators: O3, SP, VIT, SU

Makes 2 servings • Prep time: 5 minutes

A healthier and delicious twist on the classic tuna salad.

140g canned red salmon, drained
2 tablespoons avocado mayonnaise
2 tablespoons tahini
Juice of ½ fresh lemon
25g celery, chopped
½ teaspoon ground turmeric
Freshly ground black pepper
Cos lettuce leaves or mixed salad leaves, for serving
100g broccoli sprouts (see page 171)

1. In a bowl, combine the salmon, mayonnaise, tahini, lemon juice, celery and turmeric, season with pepper and mash with a fork until well combined.
2. If liked, serve the salmon salad in a lettuce wrap or over a large plate of mixed salad leaves with broccoli sprouts on top.

Nutritional analysis per serving (1 salad): fat 24g, protein 18g, carbohydrate 5g, net carbs 3g

Superfood BLT Bowls

Autophagy activators: SP, SA, PO, VIT, SU, PO

Makes 2 servings • Prep time: 10 minutes • Cook time: 5 minutes

A healthier twist on a traditional BLT, the Superfood BLT combines all the classic ingredients but gives you a boost of nutrient density. This is great to make in the morning and take to work for lunch; the hardy kale won't get soggy from the dressing.

2 tablespoons cider vinegar
2 tablespoons avocado oil
1 tablespoon Dijon mustard
1 teaspoon honey (optional)
¼ teaspoon sea salt flakes
¼ teaspoon freshly ground black pepper
4 rashers nitrite-free bacon
55g finely chopped leafy greens of choice (cos, kale, mixed salad leaves)
300g cherry tomatoes, halved
35g sunflower seeds
55g broccoli sprouts (see page 171)
1 avocado, sliced

1. In a small bowl, combine the vinegar, oil, mustard, honey, salt and pepper and use a whisk or a fork to combine well. Set aside.
2. In a medium-sized frying pan, cook the bacon over a high heat for about 3 minutes on each side, or until crispy. Transfer the bacon to a kitchen paper-lined plate to drain. Roughly chop.
3. In a large bowl, combine the leafy greens, tomatoes and sunflower seeds. Pour the dressing over the salad and mix with the greens, making sure the dressing coats everything well.
4. Scatter the bacon over the salad and top with the sprouts and avocado. Serve immediately.

Nutritional analysis per serving (½ recipe): fat 35g, protein 13g, carb 27g, net carbs 19g

Loaded Portobello Mushrooms

Autophagy activators: SP, SM, PO, VIT, SU, SA

Makes 2 servings • Prep time: 5 minutes • Cook time: 13 minutes

Black beans are packed with protein, making this a great dish for when
you're in the mood for a vegetarian dinner on a High day.

2 large portobello mushrooms
1 tablespoon avocado oil
½ yellow onion, diced
30g spinach
100g canned black beans, rinsed and drained
85g broccoli, chopped
175g guacamole (ready-prepared is fine)
30g Cheddar cheese, grated
Sea salt flakes & freshly ground black pepper

1. Preheat the oven to 190°C/Gas Mark 5.
2. Gently clean the mushrooms and trim the stalks. Drizzle with a little
    oil, season with salt and bake for 7 minutes, or until slightly softened.
    Turn the oven off when done, but keep the door shut to retain heat.
3. Meanwhile, heat the remaining oil in a frying pan over a medium
    heat. Add the onion and cook until translucent, about 5 minutes.
4. Add the spinach and cook, stirring occasionally, until wilted, about
    2 minutes. Stir in the beans and transfer to a medium-sized bowl.
5. Steam the broccoli for about 4 minutes until tender. Stir the broccoli
    into the spinach mixture,then season with salt and pepper.
6. Spread a generous amount of guacamole on the gill side of the
    mushrooms. Spoon equal amounts of the bean-vegetable mixture on
    to each mushroom and top with the cheese.
7. Return the prepared mushrooms back to the still-warm oven for 3–5
    minutes, until the cheese has melted. Serve immediately.

Nutritional analysis per serving (1 mushroom): fat 23g, protein 5g, carbohydrate
29, net carbs 22g

Steak Power Bowls

Autophagy activators: SP, SU, PO, PV, VIT, SM

Makes 2 servings • Prep time: 10 minutes • Cook time: 15 minutes

You'll love the combination of the salty feta cheese with the sweetness of the strawberries in this steak salad, a powerhouse of nutrients that makes a great meal.

2 x 115-g rump or sirloin steaks
½ teaspoon sea salt flakes
½ teaspoon freshly ground black pepper
115g mixed salad leaves
175g broccoli, finely chopped
150g strawberries, sliced
25g fresh parsley, chopped
55g broccoli sprouts (see page 171)
2 tablespoons cider vinegar
2 tablespoons avocado oil
55g feta cheese, crumbled
125g walnuts, chopped

1. Heat a grill or griddle pan to medium-high.
2. Sprinkle both sides of the steaks with the salt and pepper. Grill until golden brown on the first side, 4–5 minutes. Flip the steaks and grill for 3–5 minutes more for medium-rare, or until cooked to your liking. Leave to rest on a chopping board.
3. In a bowl, mix the salad leaves, broccoli, strawberries, parsley and broccoli sprouts. Add the vinegar and oil and toss to combine.
4. Divide the salad between two bowls. Slice the steaks across the grain into 1-cm-thick slices and place on top of the salad. Scatter some feta cheese and walnuts over each serving and serve warm or at room temperature.

Nutritional analysis per serving (½ recipe): fat 33g, protein 39g, carbohydrate 20g, net carbs 13g

Prawn and Broccoli Stir-Fry

Autophagy activators: SP, SU, SM, PO, SA, VIT, PB, O3

Makes 4 servings • Prep time: 15 minutes • Cook time: 30 minutes

450g prawns, peeled and deveined
3 spring onions, thinly sliced
2 garlic cloves, very finely chopped
1 x 2.5-cm piece fresh root ginger, peeled and very finely chopped
1 tablespoon gluten-free coconut aminos
1 teaspoon pure maple syrup (optional)
1 tablespoon dark sesame oil
3 tablespoons avocado oil
450g broccoli, cut into florets and stalks, peeled and sliced
Sea salt flakes and freshly ground black pepper
¾ teaspoon chilli flakes (optional)
2 tablespoons sesame seeds

1. In a bowl, toss the prawns, spring onions, half the garlic, half the ginger, the coconut aminos, maple syrup and sesame oil together. Cover and leave to marinate in the refrigerator for up to 4 hours.
2. In a large frying pan, heat 1 tablespoon of the avocado oil over a high heat. Add the broccoli stalks and cook, stirring constantly, for 30 seconds. Add the broccoli florets, the remaining garlic and ginger, 2 tablespoons water. Stir-fry until the broccoli is bright green, about 2 minutes. Season with salt and pepper and transfer to a plate.
3. In the same frying pan, heat the remaining 2 tablespoons avocado oil over a medium heat. Add the marinated prawns and chilli flakes. Cook, stirring continuously, for about 3 minutes. Return the broccoli to the frying pan and add more water if the pan looks dry. Stir-fry until the prawns are cooked through, 1–2 minutes more.
4. Scatter with the sesame seeds and serve immediately.

Nutritional analysis per serving (without kimchi): fat 18g, protein 20g, carbohydrate 12g, net carbs 8g

Chicken with Sautéed Red Cabbage and Rice

Autophagy activators: SA, SP, SU, SM, PO, VIT

Makes 4 servings • Prep time: 10 minutes • Cook time: 35 minutes

Black rice and red cabbage make this a polyphenol-rich, hearty dinner with a hint of sweetness from the apple. For ease and convenience, use boneless, skinless chicken thighs, since the chicken needs to be cut into cubes (in other recipes, try to include the skin and bones when cooking chicken, as they offer even more nutrients). If you have access to free-range organic chicken, then choosing fatty thigh meat with skin is optimal. Since toxins are stored in fat, you'll want to avoid fatty cuts of meat if you don't have access to organic meat. Choose less fatty breast cuts and add extra coconut oil during the cooking process.

Tip: for better glucose control, cool your rice! After the rice is cooked, allow it to cool in the refrigerator for 12 hours to form resistant starch. Consuming resistant starch improves your insulin response and makes it less likely you'll gain weight or throw your blood sugar out of kilter. You can either eat the rice cold or gently reheat it.

100g black rice
3 tablespoons plus 1 teaspoon coconut oil
1 medium onion, thinly sliced
450g chicken thighs, cut into 2.5-cm cubes
1 red cabbage, outer leaves removed, cored and very thinly sliced
1 tablespoon cider vinegar
1 bay leaf
Sea salt and freshly ground black pepper
1 apple, peeled, cored and grated

1. Cook the rice according to the instructions on the packet, but add 1 teaspoon coconut oil while cooking.
2. In a large frying pan, melt 3 tablespoons coconut oil over a medium

heat. Add the onion and cook, stirring, for about 5 minutes, until translucent.

3. Add the chicken and cook, stirring continuously, until almost cooked through, about 5 minutes.
4. Add the cabbage, vinegar, bay leaf, and 125ml water and season with salt and pepper.
5. Bring to the boil, then turn down the heat to low, cover and simmer for 20 minutes.
6. Add the apple and cook for 5 minutes more.
7. Serve the chicken and vegetables warm or cold, over the rice.

Nutritional analysis per serving (¼ recipe): fat 32g, protein 24g, carbohydrate 40g, net carbs 32g

Wild Salmon with Mixed Herby Butter

Autophagy activators: O3, SP, VIT, PO

Makes 4 servings • Prep time: 3 minutes • Cook time: 5 minutes

This great dish feels light but is actually really filling. Serve with a large green salad. Also, leftover salmon is a great protein option to use instead of canned in the Salmon Salad on page 218 for a high-protein lunch option. Simply use a fork to shred the salmon.

4 x 175-g salmon fillets
35g unsalted butter
1 lemon, halved
2 teaspoons fresh thyme, or ½ teaspoon dried
2 teaspoons chopped fresh dill, or ½ teaspoon dried
2 garlic cloves, very finely chopped
Sea salt flakes and freshly ground black pepper

1. Preheat the grill to high with a rack 15cm from the heating element. Line a baking tray with baking paper.
2. Place the salmon on the paper. Place 1 teaspoon of the butter on top of each fillet. Squeeze lemon juice over the fillets. Sprinkle evenly with the thyme, dill and garlic and season with salt and pepper.
3. Broil for 5 minutes, or until cooked to your liking, taking care not to burn the garlic.
4. Transfer the fish to individual plates. Top each fillet with one-quarter of the remaining butter and allow to fully melt. Serve warm.

Nutritional analysis per serving (one 6-ounce fillet): fat 18g, protein 37g, carbohydrate 1g

Beef Stew

Makes 4 servings • Prep time: 15 minutes • Cook time: 2 hours

This stew makes a hearty winter dinner and is a great dish to freeze for an easy lunch on High days. If using fresh tomatoes, leave the skins on to receive more of the antioxidant lycopene. Otherwise, canned tomatoes are perfectly acceptable, just opt for a BPA-free can.

2 tablespoons avocado oil
675g boneless stewing steak
1 tablespoon sea salt flakes
2 tablespoons freshly ground black pepper
1 medium onion, diced
2 carrots, chopped into large chunks
6 garlic cloves, minced
½ teaspoon dried oregano
½ teaspoon ground cumin
1 teaspoon ground turmeric
2 x 400-g cans crushed tomatoes
250ml reduced-salt beef stock
55g pitted green olives, chopped
35g capers, drained

1. Preheat the oven to 150°C/Gas Mark 2.
2. In a large ovenproof pot, heat the oil over a medium-high heat.
3. Season both sides of the beef with the salt and pepper.
4. Seal until nicely browned all over, about 2 minutes per side.
5. Add the onion, carrots, garlic, oregano, cumin, turmeric, tomatoes and stock. Cover and braise in the oven until the meat is fork-tender, about 1 hour 30 minutes.
6. Remove from the oven and stir in the olives and capers. Serve.

Nutritional analysis per serving (¼ recipe): fat 28g, protein 39g, carbohydrate 27g, net carbs 22g

Almond Flatbread

Autophagy activators: SP, SA, SU, PO, VIT

Makes 4 servings • Prep time: 5 minutes • Cook time: 25 minutes

This flatbread uses high-protein almond flour instead of wheat or other grain-based flour, giving you a bread that won't cause a spike in your blood sugar. Enjoy it with Tahini (page 260).

175g almond flour
1 teaspoon sea salt flakes
1 teaspoon freshly ground black pepper
4 tablespoons avocado oil, plus more for brushing
½ large onion, thinly sliced
70g kale, finely chopped
2 teaspoons chopped fresh rosemary

1. Preheat the oven to 230°C/Gas Mark 8. Put a heavy-based ovenproof frying pan into the oven to preheat.
2. In a large bowl, combine the almond flour, salt and pepper. While whisking, slowly add 250ml lukewarm water and whisk to eliminate lumps. Stir in 2 tablespoons of the oil. Cover and leave to sit while the oven heats, or for up to 12 hours. The mixture should have the consistency of heavy cream.
3. Carefully remove the hot pan from the oven, pour the remaining 2 tablespoons oil into the pan and swirl to coat. Add the onion and return the pan to the oven. Bake, stirring once or twice, until the onion is well browned, 6–8 minutes. Add the kale and rosemary and stir to combine.
4. Carefully remove the pan from the oven and transfer the onion-kale mixture to the bowl along with the almond flour mixture. Stir to combine, then immediately pour the mixture into the pan.
5. Bake for 10–15 minutes, until the edges look set. Remove from the oven and switch the oven to grill, with a rack a few centimetres away from the heating element.
6. Brush the top of the bread with 1 to 2 tablespoons oil. Grill just long

enough for the bread to brown and blister a little on top.

7. Cut the bread into four wedges, and serve hot or warm with some free-range ghee or butter.

Nutritional analysis per serving (¼ flatbread): fat 28g, protein 6g, carbohydrate 8g, net carbs 4g

Turkey Veggie Bake

Autophagy activators: SP, SM, VIT, PO, SA

Makes 4 servings • Prep time: 15 minutes • Cook time: 35 minutes

This is perfect 'comfort food'… make extra and store it in the freezer for a future lunch or dinner.

225g lentil pasta
4 tablespoons avocado oil
2 onions, chopped
225g turkey mince
1½ teaspoons sea salt flakes
1½ teaspoons freshly ground black pepper
2 tablespoons chopped fresh basil, or ½ teaspoon dried
4 courgettes, cut into matchsticks
140g button mushrooms, chopped
85g fresh spinach
115g full-fat mozzarella cheese, grated

1. Preheat the oven to 230°C/Gas Mark 8.
2. Cook the pasta until al dente, according to the packet instructions. Drain and set aside.
3. Meanwhile, heat 2 tablespoons oil in a frying pan over a medium heat. Add the onions and cook, stirring often, until translucent, about 5 minutes.
4. Add the turkey, salt and pepper. Cook for 5 minutes, breaking up the meat with a wooden spoon. Add the basil, courgettes, mushrooms and spinach and cook for 5 minutes more. Remove from the heat.
6. In a 23 x 33-cm baking dish, layer half the pasta, then the vegetables and turkey, then half the mozzarella. Top with the remaining pasta and scatter with the remaining mozzarella. Drizzle with the remaining oil.
7. Bake for 15 minutes, or until the pasta is golden brown and crispy.

Nutritional analysis per serving (¼ recipe): fat 38g, protein 44g, carbohydrate 26g, net carbs 14g

Herbed Pork Loin Roast

Autophagy activators: SP, PO, SP

Makes 2 servings • Prep time: 5 minutes • Cook time: 25 minutes

A delicious cut of meat, pork loin roast makes for a special dinner. The polyphenols in the avocado oil, thyme, rosemary and garlic provide antioxidant protection from compounds that are produced when meat is cooked at high temperatures. Marinating the meat for several hours enhances its flavour and antioxidant content. Serve with a large green salad or a plate of freshly vegetables and dip from the AutophaSauces section (pages 254–60). If you can't find 300-g portions, simply ask your butcher to cut a pork loin joint into 150-g portions for serving.

2 tablespoons avocado oil
1 teaspoon chopped fresh thyme
1 teaspoon chopped fresh rosemary
2 garlic cloves, very finely chopped
½ teaspoon sea salt flakes
¼ teaspoon freshly ground black pepper
1 x 300-g boneless pork loin joint
250ml beef or chicken stock

1. In a glass bowl, whisk together the oil, thyme, rosemary, garlic, salt and pepper.
2. Add the pork to the bowl and massage the pork, coating it well with the marinade. Cover and refrigerate for at least 1 hour and up to 6 hours.
3. Place the pork in a slow cooker with the bone stock and cook on low for 3–4 hours.
4. Slice the pork into 2.5-cm-thick slices and serve.

Nutritional analysis per serving (5 ounces): fat 24g, protein 33g, carbohydrate 3g, net carbs 2g

Asparagus Chips

Autophagy activators: VIT, SM, O3, PO, SP (SU as well if using cauliflower)

Makes 8 servings • Prep time: 15 minutes • Cook time: 40 minutes

Potato chips have nothing on these savoury "chips" that are equally moreish but with none of the oxidized oils and all of the energizing B vitamins and minerals to keep you healthy!

Note: Pecorino Romano cheese made from sheep's milk is less inflammatory and more easily digested than Parmesan, but with the same flavour.

> 55g grated Pecorino Romano cheese
> 4 garlic cloves, very finely chopped
> 1 teaspoon ground cumin
> 450g asparagus or 1 head cauliflower
> 1 egg
> 1 tablespoon avocado oil

1. Preheat the oven to 200°C/Gas Mark 6. Line a baking tray with baking paper.
2. Combine the cheese, garlic and ground cumin on a small flat plate. In a large mixing bowl, whisk the egg. Dredge the asparagus in the egg bowl, then roll in the cheese mixture. Place the prepared asparagus on the baking tray. Drizzle with the avocado oil.
3. Roast for 15 minutes, or until the asparagus is lightly browned and slightly crispy. Serve immediately.

Nutritional analysis per serving (⅛ recipe): fat 10g, protein 8g, carbohydrate 3g, net carbs 2g

Low-Day Recipes

Mediterranean Chopped Salad

Autophagy activators: PO, VIT

Makes 1 serving • Prep time: 5 minutes • Cook time: 5 minutes

A big salad pairs beautifully with a hearty bowl of soup. Enjoy with a Low-day side dish – I like it with Lemon Asparagus Soup (page 244).

½ **head cos lettuce, chopped**
½ **avocado, diced**
25g pitted black olives, chopped
1 tablespoon extra virgin olive oil
1 tablespoon fresh lemon juice
2 teaspoons chopped fresh oregano, or ¼ teaspoon dried
Sea salt flakes

1. Combine the lettuce, avocado and olives in a medium-sized bowl.
2. In a small bowl, whisk together the oil, lemon juice and oregano and season with salt.
3. Drizzle the oil mixture over the vegetables and toss to combine. Season with salt and serve immediately.

Nutritional analysis per serving (1 salad): fat 29g, protein 5g, carbohydrate 16g, net carbs 5g

Grapefruit Salad

Autophagy activators: VIT, PO, SP

Makes 1 serving • Prep time: 5 minutes • Cook time: 5 minutes

Grapefruit is a wonderful fruit to eat alone, but it pairs beautifully with the buttery smoothness of avocado and subtle warmth of raw ginger. Enjoy with a Low-day side dish – I think it pairs well with a Veggie Kebab (page 248).

1 small grapefruit, peeled and segmented
25g rocket
55g cos or Little Gem lettuce, chopped
¼ teaspoon finely chopped fresh ginger
1 tablespoon avocado oil
½ avocado, sliced
4 tablespoons chopped fresh coriander

1. Combine the grapefruit, rocket, cos and ginger in a bowl.
2. Drizzle with the oil and toss to coat.
3. Top with the avocado and coriander and enjoy immediately.

Nutritional analysis per serving (1 salad): fat 24g, protein 4g, carbohydrate 25g, net carbs 17g

Cucumber Salad

Autophagy activators: PO, SU, VIT, SP

Makes 1 serving • Prep time: 5 minutes • Cook time: 5 minutes

A crisp salad perfect for a hot summer day. Make and serve immediately, or chill without the avocado and top with the avocado just before serving. Enjoy with a Low-day side dish like Sautéed Broccoli (page 243).

1 tablespoon fresh lemon juice
1 tablespoon avocado oil
1 cucumber, sliced
100g red cabbage, shredded
2 tablespoons chopped fresh dill
½ avocado, sliced

1. In a medium-sized bowl, combine the lemon juice and oil and whisk well.
2. Add the cucumber, cabbage and dill and toss to coat.
3. Top with the avocado and enjoy immediately.

Nutritional analysis per serving (1 salad): fat 24g, protein 5g, carbohydrate 18g, net carbs 7g

Veggie Roast

Autophagy activators: SU, SM, VIT, PO, SP

Makes 2 servings • Prep time: 10 minutes • Cook time: 25 minutes

I love the versatility of this veggie roast because you can make it for dinner and then enjoy it for your next High-day lunch by simply adding a protein. Plus, roasted veggies are delicious with one of the AutophaSauces on pages 254–60. On a Low day, I think this pairs well with Spaghetti Squash (page 247).

1 medium courgette, halved lengthways and sliced into
5-mm-thick half-moons
½ head cauliflower, diced
2 tablespoons sunflower seeds
4 tablespoons avocado oil
½ teaspoon ground turmeric
Freshly ground black pepper
55g broccoli sprouts (see page 171)

1. Preheat the oven to 230°C/Gas Mark 8.
2. In a large bowl, combine the courgette, cauliflower and sunflower seeds. Add the oil and turmeric, season with pepper and toss to coat.
3. Spread the vegetables in a baking tray. Roast for 15–20 minutes, until the vegetables are tender and lightly golden.
4. Top with the sprouts and serve.

Nutritional analysis per serving (½ recipe): fat 32g, protein 6g, carbohydrate 12g, net carbs 7g

Broccoli Soup

Autophagy activators: VIT, PO, SU, SP, SM

Makes 4 servings • Prep time: 5 minutes • Cook time: 30 minutes

You'll love the velvety texture and brilliant green colour of this puréed broccoli soup. Make a double batch and freeze for later use. Serve with a large mixed green salad or a generous helping of any vegetable of your choice.

2 heads broccoli
1 tablespoon ghee or avocado oil
1 shallot, very finely chopped
2 cloves garlic, very finely chopped
1 avocado, sliced
125ml full-fat coconut milk
350ml bone stock or stock of your choice
¾ teaspoon sea salt flakes
1 teaspoon black pepper
20g fresh basil or dill

1. Roughly chop the broccoli florets and stalks (discard the tough lower stalk). Place in a steam basket and lightly cook until fork tender, about 10 minutes.
2. Over a medium heat, heat the ghee or oil and fry the shallot and garlic. Stir continuously to avoid burning.
3. Combine the broccoli and the shallot mixture in a high-speed blender or food processor with the avocado, coconut milk, bone stock, salt and pepper, and process until smooth.
4. Garnish with fresh herbs and additional avocado slices.
5. If you prefer having your soup hotter, reheat on the hob at a gentle simmer for 2–3 minutes.

Nutritional analysis per serving (¼ recipe): fat 14g, protein 7g, carbohydrates 10g, net carbs 6g

Pesto Courgetti

Autophagy activators: PO, SU, SA, SP

Makes 2 servings • Prep time: 15 minutes • Cook time: 20 minutes

Courgette makes a perfect substitute for pasta when you want more nutrients and fewer processed carbohydrates. Adding Greens Pesto gives it a boost in flavour and polyphenols! Serve with a large mixed green salad or a bowl of Lemon Asparagus Soup (page 244).

2 tablespoons avocado oil
2 garlic cloves, minced
70g kale, chopped
1 medium courgette, spiralized or sliced with a vegetable peeler
into long strands (aka courgetti)
4 tablespoons Greens Pesto (page 257)

1. In a large frying pan, heat the oil over a medium heat. Add the garlic and cook, stirring frequently, until fragrant.
2. Add the kale and cook, stirring, until it begins to wilt.
3. Add the courgetti and cook until tender, about 5 minutes.
4. Turn down the heat to low. Add the pesto and cook, stirring, until the courgetti are evenly coated.
5. Divide the courgetti between two bowls and enjoy immediately.

Nutritional analysis per serving (½ recipe): fat 28g, protein 2g, carbohydrate 7g, net carbs 6g

Powerphenol Purple Smoothie

Autophagy activators: VIT, PO, SU, SA

Makes 2 servings • Prep time: 3 minutes • Cook time: 1 minute

This makes a convenient and nutrient-packed lunch when you're on the go. Enjoy with a big plate of fresh raw veggies with tahini on the side. Or chill your smoothie in the refrigerator for a few hours and have it as a late-afternoon snack.

70g frozen blackberries
70g frozen blueberries
70g coarsely chopped kale leaves
250ml unsweetened almond milk
2 tablespoons coconut butter
1 tablespoon raw cacao powder
1 tablespoon hemp seeds, chia seeds or ground linseed
¼ avocado
Ice cubes, as desired

Combine all the ingredients in a high-speed blender and blend well for 20–30 seconds. Pour into a glass and enjoy, or cover and refrigerate for up to a few hours.

Nutritional analysis per serving (½ smoothie): fat 20g, protein 5g, carbohydrate 20g, net carbs 13g

Cauliflower Mushroom Roast

Autophagy activators: VIT, SM, SP, SU, PO

Makes 4 servings • Prep time: 10 minutes • Cook time: 55 minutes

Hearty shiitake and chestnut mushrooms complement any meal on a Low day when you want something 'meaty', filling and satisfying. Serve alongside some Spaghetti Squash (page 247).

1 head cauliflower, cut into small florets
2 tablespoons avocado oil
1 tablespoon chopped fresh thyme, or ½ teaspoon dried
¼ teaspoon turmeric
60g shiitake mushrooms, cut into 5-mm slices
60g chestnut mushrooms, cut into 5-mm slices
2 tablespoons free-range ghee or butter, melted
1 teaspoon sea salt flakes
Freshly ground black pepper
20g Pecorino Romano cheese, grated

1. Preheat the oven to 230°C/Gas Mark 8.
2. In a bowl, combine the cauliflower, 1 tablespoon of the oil, the thyme and the turmeric. Rub the seasonings and oil into the cauliflower. Transfer to a baking tray and roast for 20 minutes.
3. In the same bowl, combine the mushrooms, butter or ghee, and remaining 1 tablespoon oil and rub to coat. Add to the baking tray with the cauliflower and roast for 10–15 minutes more, until the cauliflower is golden brown.
4. Remove from the oven, scatter with the salt, pepper to taste and cheese, and serve immediately.

Nutritional analysis per serving (¼ recipe): fat 18g, protein 8g, carbohydrate 8g, net carbs 5g

Taco Bowl Salad

Autophagy activators: PO, VIT, SA, SP

Makes 1 serving • Prep time: 5 minutes • Cook time: 5 minutes

Serve this salad bowl alongside a Veggie Kebab (page 248). For more flavour and polyphenol content, top with Double Greens Dip (page 255). Also, this simple taco bowl salad can easily be eaten on a High day if you add chicken, beef or pork.

55g mixed salad leaves
40g cooked black beans
½ large avocado, chopped
2 tablespoons ready-prepared salsa
4 tablespoons chopped fresh coriander
¼ teaspoon ground cumin
Fresh lime juice
Sea salt flakes
25g broccoli sprouts (see page 171)

1. In a bowl, combine the mixed salad leaves, beans and avocado.
2. Add the salsa and coriander and toss to combine.
3. Season with the cumin. Squeeze fresh lime juice over the top and season with salt. Top with the broccoli sprouts and serve.

Nutritional analysis per serving (1 salad): fat 16g, protein 9g, carbohydrate 26g, net carbs 9g

Greens Bowls

Makes 2 servings • Prep time: 5 minutes • Cook time: 5 minutes

Eating a variety of raw and cooked greens daily is so important. Try this dish when you want a Japanese flavour. Enjoy with a serving of Cauliflower Rice (page 245) and Miso-Glazed Courgette (page 246) for a satisfying Orient-inspired low-protein meal that will entertain your taste buds. On a High day, this bowl tastes great with grains like quinoa and rice or with salmon or beef.

2 tablespoons coconut oil or expeller-pressed unrefined sesame oil
2 garlic cloves, very finely chopped
1 tablespoon gluten-free tamari or coconut aminos
70g kale, chopped
55g spinach
2 tablespoons sesame seeds
2 teaspoons dulse flakes, soaked and drained according to packet
instructions (optional)

1. In a medium-sized frying pan, melt the coconut oil over a medium-high heat. Add the garlic and cook, stirring constantly, for 1 minute. Add the tamari, kale and spinach and cook until wilted.
2. Divide the greens between two bowls, garnish with the sesame seeds and dulse flakes (if using) and serve.

Nutritional analysis per serving (½ recipe): fat 17g, protein 4g, carbohydrate 4g, net carbs 3g

Honey-Roasted Brussels Sprouts

Autophagy activators: SU, SP, PO

Makes 2 servings • Prep time: 15 minutes • Cook time: 5 minutes

When you cook cruciferous veggies slow and low, their natural sweetness comes out. In this dish, you get that same deep flavour without the time! With just a touch of honey, you can recreate the deep flavour of a slow-roasted vegetable in only minutes. The rich taste of cooked honey turns this dish into something mildly decadent! With the polyphenols and high smoke point of avocado oil, you can feel good about eating this crispy treat. Try serving this with Miso-Glazed Courgette (page 246).

175g Brussels sprouts, halved
85g broccoli, chopped
1 teaspoon honey
3 tablespoons avocado oil
1 teaspoon sea salt flakes
1 teaspoon freshly ground black pepper

1. In a large bowl, combine the Brussels sprouts, broccoli, honey, 1 tablespoon of the oil, the salt and pepper.
2. Heat a large heavy-based frying pan over a high heat for 1–2 minutes. Add the remaining 2 tablespoons oil and reduce the heat to medium. Add the Brussels sprouts and broccoli and stir to combine. Cook for about 3 minutes, allowing the vegetables to brown before flipping to get good caramelization on both sides.
3. Remove from the heat and divide between two bowls. Serve hot.

Nutritional analysis per serving (½ recipe): fat 21g, protein 4g, carbohydrates 14g, net carbs 10g

Sautéed Broccoli

Autophagy activators: SU, SM, PO, SA, SP

Makes 4 servings • Prep time: 10 minutes • Cook time: 10 minutes

Add sprouts to this simple yet delicious dish to increase its sulforaphane content. This makes a great addition to any meal.

3 tablespoons avocado oil
3 garlic cloves, chopped
1 head broccoli, trimmed and cut into bite-sized pieces
125ml vegetable stock
½ teaspoon sea salt flakes
Zest of 1 lemon
1 tablespoon fresh lemon juice
55g broccoli sprouts (see page 171)
1 large avocado, sliced

1. In a large frying pan, heat the oil over a medium heat. Add the garlic and cook, stirring continuously, until fragrant, about 30 seconds. Stir in the broccoli and cook until the broccoli is bright green, about 3 minutes.
2. Add the stock and season with the salt, lemon zest and lemon juice. Cook until the broccoli is tender, 3–5 minutes.
3. Top with the sprouts and avocado and serve.

Nutritional analysis per serving (¼ recipe): fat 16g, protein 5g, carbohydrate 15g, net carbs 9g

Lemon Asparagus Soup

Autophagy activators: PO, VIT, SP, SM, SA

Makes 4 servings • Prep time: 10 minutes • Cook time: 50 minutes

A delicious soup for spring and early summer. Make a double batch and freeze half for later. Serve with a large mixed green salad or as a complement to a Mediterranean Chopped Salad (page 232). You can also use homemade or ready-prepared bone stock in place of the vegetable stock on a High day.

3 tablespoons avocado oil
2 medium brown onions, finely chopped
2 garlic cloves, very finely chopped
1 litre vegetable stock
800g asparagus, tips removed, stalks cut into 1-cm pieces
1 teaspoon sea salt flakes
1 tablespoon chopped fresh dill, or ½ teaspoon dried
2 tablespoons fresh lemon juice
20g Pecorino Romano cheese grated

1. In a large pot, heat the oil over a medium heat. Add the onions and cook for about 5 minutes. Add the garlic and cook, stirring, until the onions are translucent, about 3 minutes more.
2. Add the stock and bring to a low boil. Add the asparagus, salt and dill and bring back to the boil. Reduce the heat to low and simmer for 20–30 minutes until you can easily pierce the asparagus with a fork. Remove from the heat.
3. Using a stick blender, blend the soup directly in the pot until smooth. Strain the soup through a fine-mesh sieve into a large bowl. (Use caution while transferring the hot liquid.)
4. Pour the soup back into the pot and return it to a medium heat. Add the lemon juice and Pecorino Romano and heat until hot all the way through, 1–2 minutes. Serve immediately.

Nutritional analysis per serving (¼ recipe): fat 14g, protein 9g, carbohydrate 16g, net carbs 12g

Cauliflower Rice

Autophagy activators: SU, SM, PO

Makes 2 servings • Prep time: 20 minutes • Cook time: 10 minutes

"Ricing" your cauliflower is a creative way to sneak in more vegetables while keeping excess starch and protein out. Change it up by adding different spices to season the "rice", such as cumin and paprika for a Mexican flair or turmeric and garam masala for an Indian twist. The sky's the limit, so don't be afraid to experiment with your favourite flavours. I like to pair my Cauliflower Rice with a Greens Bowl (page 241).

1 head cauliflower, quartered, stalks chopped
2 tablespoons coconut oil
1 onion, chopped
1 garlic clove, very finely chopped
Sea salt flakes and freshly ground black pepper

1. Pat the cauliflower dry and put it into a food processor. Pulse until broken down to the size of rice grains; do not over-process or the cauliflower will become mushy.
2. In a large frying pan, melt the coconut oil over a medium heat. Add the onion and garlic and cook, stirring, for 3–4 minutes, until the onion is translucent.
3. Add the cauliflower and cook, stirring, for 4–5 minutes, until tender but not mushy.
4. Season with salt and pepper and serve.

Nutritional analysis per serving (½ recipe): fat 13g, protein 7g, carbohydrate 22g, net carbs 13g

Miso-Glazed Courgette

Autophagy activators: SA, SM, PB, PO

Makes 2 servings • Prep time: 5 minutes • Cook time: 15 minutes

Due to its high water content, courgette is a great sponge for soaking up the richness of spices and sauces. In this dish the savoury, salty miso deepens the flavour profile, making the courgette a memorable delicacy you will return to time and time again. Try pairing this dish with Honey-roasted Brussels Sprouts (page 242).

1½ tablespoons miso paste
1½ teaspoons mirin
1 medium courgette, trimmed and halved lengthways
2 teaspoons toasted sesame oil
1 tablespoon chopped fresh basil
1 tablespoon sesame seeds
55g kimchi

1. Preheat the oven to 200°C/Gas Mark 6.
2. In a small bowl, whisk together the miso, mirin and 2 tablespoons water until combined.
3. Brush the cut sides of the courgette with the oil. Roast the courgette for 10 minutes. Brush the miso mixture on top of each courgette half and roast for another 10 minutes, or until golden brown.
4. Transfer to a serving platter and scatter over the basil and sesame seeds. Serve with the kimchi on the side.

Nutritional analysis per serving (½ recipe): fat 8g, protein 5g, carbohydrate 12g, net carbs 9g

Spaghetti Squash

Autophagy activators: VIT, SM, PO, SU, SP

Makes 2 servings • Prep time: 15 minutes • Cook time: 1 hour

You'll save on refined carbs but gain antioxidants and amazing flavour by trading in traditional noodles for spaghetti squash. Spaghetti squash is a versatile fruit (yes, it's really a fruit!) and is delicious cooked and eaten with butter, or like this, with red cabbage and onion. Serve as a base for Cauliflower Mushroom Roast (page 239) or with a large mixed green salad.

> **1 medium spaghetti squash, halved and seeded**
> **2 tablespoons avocado oil**
> **½ onion, diced**
> **100g red cabbage, shredded**
> **2 tablespoons red wine vinegar**
> **2 teaspoons sea salt flakes**
> **250ml vegetable stock**

1. Preheat the oven to 200°C/Gas Mark 6.
2. Place the squash cut-side up in a roasting tin and roast for 30–40 minutes until you can easily pierce the flesh with a fork. Remove from the oven and leave until cool enough to handle.
3. Use a fork to scrape the squash flesh into long, thin strands to make 'spaghetti'. Set aside. Discard the skins.
4. In a large pan, heat the oil over a medium heat. Add the onion and cook until translucent, about 5 minutes.
5. Add the cabbage, vinegar and salt and cook until the cabbage wilts, 3–5 minutes.
6. Add the stock and simmer for 5 minutes.
7. Add the 'spaghetti' to the pan, stir to combine and cook for 2 minutes more, or until the vegetables are nice and soft without becoming mushy. Serve immediately.

Nutritional analysis per serving (½ recipe): fat 14g, protein 2g, carbohydrate 18g, net carbs 15g

Veggie Kebabs

Autophagy activators: SA, SM, SP, PO

Makes 3 servings (6 kebabs) • Prep time: 10 minutes • Cook time:
15 minutes

Enjoy this mixed-veggie kebab with any dish, any time! These kebabs are
perfectly compatible with any of the sauces or dips in the AutophaSauces
section (pages 254–60). You can buy skewers at most supermarkets.
Wooden or bamboo ones are excellent; just make sure to soak them in
water for 30 minutes before using so they don't burn when cooking.

2 small courgettes, cut crossways into 2-cm-thick slices
2 red peppers, cut into 2.5-cm pieces
30g chestnut mushrooms, sliced
3 tablespoons avocado oil
1 teaspoon sea salt flakes
1 teaspoon freshly ground black pepper
Dried oregano (optional)
Garlic powder (optional)
55g raw sauerkraut, ready-prepared or homemade (see page 171)

1. If using wooden or bamboo skewers, soak them in water to cover for
    30 minutes (see Note).
2. Preheat the grill to high.
3. In a large bowl, combine the courgetes, red peppers, mushrooms,
    oil, salt and black pepper. Season with oregano and/or garlic
    powder, if liked. Toss well to coat the vegetables.
4. Thread the vegetables on to the skewers in any order you like.
5. Grill until the vegetables start to brown, 8–10 minutes, turning the
    kebabs once halfway through the cooking time.
6. Serve with sauerkraut on the side.

Nutritional analysis per serving (2 kebabs): fat 14g, protein 2g, carbohydrate 9g,
net carbs 7g

Slow-roasted Cherry Tomatoes

Autophagy activators: SA, PO, VIT, SP

Makes 2 servings • Prep time: 15 minutes • Cook time: 3 hours

These take a bit of time in the oven, but once you taste their flavour, you'll understand why it's worth it. Perfect as a snack on their own, they also make an impressive accompaniment to other starters like hummus and veggies.

450g mixed cherry tomatoes, halved lengthways
8 garlic cloves, unpeeled
2 sprigs fresh thyme, leaves only
2 tablespoons avocado oil
¼ teaspoon sea salt flakes

1. Preheat the oven to 110°C/Gas Mark ¼. Line a baking tray with baking paper.
2. Place the tomatoes on the baking tray, cut-side up, and scatter the unpeeled garlic cloves and thyme leaves around the tray.
3. Drizzle the tomatoes and garlic with the oil and season evenly with the salt.
4. Roast for about 3 hours, until the tomatoes look shrivelled but still have some juice left.
5. Remove the tomatoes from the baking tray and serve.

Nutritional analysis per serving (½ recipe): fat 14g, protein 3g, carbohydrate 23g, net carbs 20g.

Salt-and-Vinegar Kale Crisps

Autophagy activators: SU, VIT, PO, SP

Makes 3 servings • Prep time: 15 minutes • Cook time: 15 minutes

You may be surprised by how kale can satisfy your desire for a salty crisp. Cook the leaves until just crispy but not burnt to keep the nutrients and polyphenols in the avocado oil intact and beneficial.

450g curly kale
2 tablespoons avocado oil
2 tablespoons cider vinegar
1 teaspoon coarse sea salt flakes

1. Preheat the oven to 180°C/Gas Mark 4. Line a baking tray with baking paper.
2. Using kitchen shears or your hands, separate the kale leaves from the thick stalks; discard the stalks. Cut the leaves up into pieces around 5cm long.
3. Wash and dry the kale well.
4. In a large bowl, combine the kale, oil, vinegar and salt. Using clean hands, massage the kale for 2 minutes to help break it down a bit; it should be slightly soft at this point.
5. Spread the kale into a single layer on the prepared baking tray. (Use a second baking tray if you run out of room.)
6. Bake for 5 minutes, keeping an eye on the kale so it doesn't burn. Toss with a wooden spoon and bake for 5–10 minutes more, until crispy and browned around the edges.
7. Serve immediately.

Nutritional analysis per serving (⅓ recipe): fat 9g, protein 1g, carbohydrate 3g

Nutty Chia Clusters

Autophagy activators: SA, SP, VIT, PO, O3, CA, PB

Makes 6 servings • Prep time: 20 minutes • Cook time: 1 hour cooling time

When you're craving something a little sweet on a High day, these make a delicious treat, and you get to enjoy the polyphenols from some dark chocolate.

2 tablespoons coconut oil
2 tablespoons coconut butter
1 teaspoon Ceylon cinnamon
1 serving high-quality hydrolyzed collagen peptides
(check the packet for serving size)
2 tablespoons unsweetened peanut butter or nut butter of your choice
70g desiccated coconut
25g chopped walnuts
2 tablespoons chia seeds
2 tablespoons high-quality dark chocolate chips (at least 70% cacao)

1. In a large saucepan, melt the coconut oil and coconut butter over a low heat. Add the cinnamon, collagen and nut butter and stir until smooth.
2. Remove from the heat and stir in the coconut, walnuts, chia seeds and chocolate chips until well combined.
3. Using a tablespoon or a small ice cream scoop, scoop mounds of the mixture on to a baking tray lined with baking or greaseproof paper.
4. Refrigerate until hardened, at least an hour or two. At room temperature, coconut becomes liquid, so store in a glass container in the refrigerator up to five days, or keep them in the freezer up to four months.

Nutritional analysis per serving (⅙ recipe): fat 23g, protein 7g, carbohydrate 12g, net carbs 6g

Strawberry Coconut Balls

Autophagy activators: PO, O3

Makes 16 servings • Prep time: 20 minutes • Chill time: 30 to 60 minutes

Strawberries are low in carbohydrates, high in fibre and rich in blood sugar-stabilizing polyphenols. Paired with the slightly sweet flavour coconut butter imparts, these treats make a luscious, melt-in-your mouth dessert or quick snack. Start out with one ball and if after 15 minutes you feel like eating another one would benefit you, go for it. The beauty of these is that a little can go a long way, so play around with what your body needs for that particular day.

Desiccated coconut
250g coconut butter
225g coconut oil
80g frozen or fresh strawberries
1 teaspoon monk fruit sweetener (optional)
1 teaspoon real vanilla powder (optional)

1. Place the desiccated coconut on a plate and set aside.
2. Place the coconut butter, coconut oil, strawberries, monk fruit and vanilla powder in a food processor and blend until creamy and smooth, about 3 minutes.
3. Cool the mixture in the refrigerator for 30 minutes, or until the mixture thickens.
4. Spoon out about 2 tablespoons at a time and shape into balls. Roll in coconut to cover. Chill in the refrigerator in a glass container. Best enjoyed cold.

Nutritional analysis per serving (1 ball): fat 26g, protein 1g, carbohydrate 5g, net carbs 4g

Macadamia Chocolate Bark

Autophagy activators: PO, CA, PB

Makes 16 servings • Prep time: 25 minutes • Cook time: 2 hours cooling time

Chocolate bark is a delicious way to indulge. Macadamia nuts add a great creamy flavour and some healthy monounsaturated fats. To spice up your chocolate bark even more, try a pinch of cayenne, or ½ teaspoon curry powder or ground ginger.

300g plain chocolate (at least 80% cacao), chopped
150g unsalted macadamia nuts, finely chopped

1. Line a 23-cm square baking tin with baking paper, leaving a few centimeters overhanging on two sides as handles. Fill a small saucepan with a couple centimeters of water and bring to a simmer over a medium heat.
2. Put the chocolate into a heatproof bowl that will comfortably sit on top of the saucepan without touching the water.
3. Turn off the heat, set the bowl of chocolate over the water and stir until melted and smooth. Take care not to scorch the chocolate.
4. Pour two-thirds of the melted chocolate into the lined baking tin and spread it evenly with a slotted turner. Scatter the macadamia nuts evenly over the chocolate and cover with the remaining chocolate, carefully spreading it into an even layer.
5. Refrigerate for at least 2 hours until firmly set.
6. Use the baking paper to remove the bark from the tin, and cut it into 16 squares. Store in an airtight container at room temperature.

Nutritional analysis per serving (1 square): fat 14g, protein 2g, carbohydrate 10g, net carbs 9g

Autophasauces

Not only are sauces a way to boost an ordinary meal to flavourful heights, they can also save you time (if you make extra for the next day) and can be a source of potent autophagy activators. Once you have made your own sauces, dips and dressings (many dips and sauces can be diluted to become salad dressing), you won't want to go back to ready-prepared products – they don't taste as good, and the ingredients they use more often than not contribute to inflammation, oxidation and ageing. These AutophaSauces are excellent for High or Low days or as something to dip a snack into.

Goddess Dip

Autophagy activators: PO, VIT

Makes 8 servings (about 250ml/9fl oz) • Prep time: 5 minutes

Glow like a goddess with my favourite dip made with my favourite youth-boosting actives. High in polyphenols and full of good fat, it's an easy-to-make treat to add to your table.

2 tablespoons raw cider vinegar
4 tablespoons avocado oil
½ small avocado
4 tablespoons fresh basil
4 tablespoons fresh parsley
2 tablespoons packed fresh mint
1 teaspoon monk fruit
1 teaspoon dulse flakes
Sea salt flakes and black pepper to taste

Combine all the ingredients in a food processor and process until smooth.

Nutritional analysis per serving (2 tablespoons): fat 8g, protein 0g, carbohydrate 2g, net carbs 1g

Double Greens Dip

Autophagy activators: PO, VIT

Makes 8 servings (about 250ml/9fl oz) • Prep time: 5 minutes

This greens dip will add a splash of colour and a healthy dose of antioxidants and vitamins to any fish, meat or chicken. Alternatively, it makes a delicious dip for raw carrot and celery sticks.

85g fresh basil, stalks removed
115g parsley, ends trimmed
115g avocado oil
Zest and juice of ½ lime
4 tablespoons coconut aminos

Combine all the ingredients in a food processor and process until smooth. Store in an airtight container in the refrigerator for up to 1 week.

Nutritional analysis per serving (2 tablespoons): fat 27g, protein 2g, carbohydrate 6g, net carbs 5g

Hollandaise Superfood Plus

Autophagy activators: SP, PO, VIT

Makes 8 servings (about 250ml/9fl oz) • Prep time: 5 minutes

Boost regular hollandaise sauce to superfood status when you combine vitamin K_2-rich egg yolks from free-range hens and butter or ghee (clarified butter) with the anti-inflammatory and anti-ageing spices turmeric and thyme. Enjoy on steamed asparagus, broccoli or any green vegetable.

115g unsalted butter or ghee (clarified butter)
2 teaspoons dried thyme
Zest of 1 lemon
4 teaspoons fresh lemon juice
6 egg yolks
1 teaspoon ground turmeric
Pinch of sea salt flakes, or more to taste
Pinch of freshly ground black pepper, or more to taste

1. In a small saucepan, melt the butter over a low heat. Add the thyme and lemon zest and remove from the heat.
2. Put the lemon juice, egg yolks, turmeric, salt and pepper into a blender. Blend on high for 15 seconds until thoroughly mixed.
3. With the blender running, slowly pour in the melted butter-herb mixture through the lid vent (have a towel nearby in case of splatters). Add 2 teaspoons room-temperature water and blend for 2 seconds more, until smooth and creamy. Taste and season with more salt and pepper if needed. Store in an airtight container in the refrigerator for up to 1 week.

Nutritional analysis per serving (2 tablespoons): fat 15g, protein 2g, carbohydrate 1g

Greens Pesto

Autophagy Activators: PO, VIT, SA

Makes 8 servings (about 250ml/9fl oz) • Prep time: 5 minutes

I promise you'll want to double-dip your veggie sticks into this variation of a traditional pesto. Enjoy this brilliant green, polyphenol-rich pesto raw, or spread some on meat or fish before cooking, since the avocado oil, with its high smoke point, makes it a great choice for higher-heat cooking.

115g parsley, ends trimmed
2 tablespoons fresh thyme leaves
2 teaspoons fresh rosemary leaves
4 garlic cloves, roughly chopped
100g avocado oil
Juice of 1 lemon
Pinch of sea salt flakes, or more to taste

Combine all the ingredients in a food processor and process until smooth. Taste and season with more salt if needed. Store in an airtight container in the refrigerator for up to 1 week.

Nutritional analysis per serving (2 tablespoons): fat 14g, protein 1g, carbohydrate 3g, net carbs 2g

Sun-Dried Tomato and Olive Spread

Autophagy Activators: PO, VIT, SA

Makes 8 servings (about 250ml/9fl oz) • Prep time: 5 minutes

The savoury-salty combination of this spread hits the spot when you need something that feels rich but not overwhelmingly heavy. Wrap it up in lettuce wrap or use as a dip for fresh raw vegetables. Or spread it over grilled chicken or fish.

150g macadamia nuts
55g sun-dried tomatoes
25g pitted oil-cured black or Kalamata olives
Juice of ½ lemon
¼ teaspoon sea salt flakes
Pinch of freshly ground black pepper
25g chopped fresh parsley

Combine all the ingredients in a blender and blend until smooth. Store in an airtight container in the refrigerator for up to 1 week.

Nutritional analysis per serving (2 tablespoons): fat 14g, protein 2g, carbohydrate 5g, net carbs 3g

Almond Miso Dip

Autophagy Activators: PO, VIT, PB

Makes 8 servings (about 250ml/9fl oz) • Prep time: 5 minutes

This dressing can be used to coat courgette, carrot, sweet potato or beetroot noodles to give your dish that perfect tangy Orient flair. The almonds contain healthy monounsaturated fats to help you absorb the vitamin A in your vegetables. Enjoy cold for best flavour!

125g unsweetened almond butter
2 tablespoons white miso paste
55g carro,t grated
Juice of ½ lime
1 tablespoon coconut aminos
1 x 2.5-cm piece fresh root ginger, peeled and grated
¼ teaspoon sea salt flakes
Pinch of freshly ground black pepper
4 tablespoons chopped fresh coriander

Combine all the ingredients and 4 tablespoons water in a blender and blend until smooth. Store in an airtight container in the refrigerator for up to 1 week.

Nutritional analysis per serving (2 tablespoons): fat 9g, protein 4g, carbohydrate 6g, net carbs 3g

Maple-Scented Tahini Dip

Autophagy Activators: PO, VIT

Makes 8 servings (about 250ml/9fl oz) • Prep time: 5 minutes

Use this as a decadent dressing over hearty greens like a shredded kale salad or as a dip for freshly cut vegetables. The flavour is a mixture between earthy nuttiness from the tahini and sweetness from the maple. You will be licking your plate to get the last bit!

160g tahini
1 tablespoon avocado oil
2 teaspoons pure maple syrup
Juice of 1 lemon
2 small garlic cloves
½ teaspoon sea salt flakes
¼ teaspoon freshly ground black pepper

Combine all the ingredients and 125ml water in a blender and blend until smooth. Store in an airtight container in the refrigerator for up to 1 week.

Nutritional analysis per serving (2 tablespoons): fat 11g, protein 4g, carbohydrate 3g, net carbs 0g

Protein Content

Use the chart below to determine the protein content of various foods.

FOOD	PORTION SIZE	PROTEIN (IN GRAMS)
Almond butter	25g	6g
Almonds	2 tablespoons	4g
Amaranth	55g, uncooked	8g
Baked beans	60g	3g
Baked potato	medium-sized	3g
Barley, hulled	60g, uncooked	5g
Beef fillet	85g	24g
Black beans	40g	4g
Broccoli	90g	2g
Brussels sprouts	90g	3g
Buckwheat	40g, uncooked	6g
Bulgur	50g, uncooked	6g
Cannellini beans	40g	4g
Cashews	25g	4g
Chia seeds	25g	5g
Chicken breast, boneless, skinless	115g	34g
Chickpeas	40g	3g
Cod	115g	24g
Cottage cheese	115g	14g
Edamame	40g	4g
Egg	1 medium	6g
Egg white	1 medium	4g
Greek yogurt, whole milk	225g	20g
Gruyère cheese	25g	8g
Halibut	115g	26g
Hemp seeds	25g	6g
Hummus	2 tablespoons	1g

FOOD	PORTION SIZE	PROTEIN (IN GRAMS)
Khorasan	40g, uncooked	7g
Kidney beans	40g	4g
Lentils	50g	4g
Lobster	85g	16g
Mozzarella (reduced fat)	25g	7g
Oats, rolled	40g, uncooked	7g
Peanut butter	1 tablespoon	7g
Pinto beans	40g	5g
Pistachios	25g	6g
Pork chop	115g	31g
Prawns	115g	28g
Pumpkin seeds	6g	25g
Quinoa	40g, uncooked	7g
Salmon	115g	24g
Scallops	85g	14g
Sea bass	115g	26g
Soya milk	250ml	8g
Soya beans	40g, cooked	15g
Spelt	40g, raw	6g
Spinach	25g	3g
Split peas	50g	4g
Steak mince	115g	29g
Sun-dried tomatoes	25g	4g
Sunflower seeds, with hulls	25g	6g
Sweetcorn	70g	3g
Tempeh	115g	21g

FOOD	PORTION SIZE	PROTEIN (IN GRAMS)
Tofu	115g	7g
Trout	115g	28g
Tuna, canned	115g	29g
Turkey, roasted	115g	34g
Walnuts	25g	4g
Wheatberries	50g, uncooked	7g

Chapter 13

Glow15 Workouts

HIIT and RET

All the following workouts can be done at home or in a gym.

There is no need to purchase any additional equipment.

Each workout is 30 minutes long and can be modified for beginner and advanced fitness levels.

High-Intensity Interval Training (HIIT) Workouts

Option 1: On the Glow with HIIT (30 Minutes)

Choose whatever cardiovascular activity you like:

- Walking
- Jogging
- Swimming
- Cycling
- Rowing
- Running

For Beginners

- 0 to 5 minutes: Warm up doing your chosen cardiovascular activity (perceived effort of 3 to 4 on a scale of 1 to 10).
- 6 to 25 minutes: Do 30 seconds of hard effort (7 or 8 on the scale), followed by 30 seconds of lighter effort (4 or 5 on the scale).
- 26 to 30 minutes: Cool down (3 or 4 on the scale).

For Intermediate Levels

- 0 to 5 minutes: Warm up doing your chosen cardiovascular activity (perceived effort of 3 to 4 on a scale of 1 to 10).
- 6 to 25 minutes: Do 1 minute of hard effort (8 or 9 on the scale), followed by 45 seconds of lighter effort (4 or 5 on the scale).
- 26 to 30 minutes: Cool down (about 3 on the scale).

For Advanced Levels

- 0 to 5 minutes: Warm up doing your chosen cardiovascular activity (perceived effort of 3 to 4 on a scale of 1 to 10).
- 6 to 25 minutes: Do 2 minutes of hard effort (7 or 8 on the scale), followed by 45 seconds of lighter effort (4 or 5 on the scale).
- 26 to 30 minutes: Cool down (about 3 on the scale).

Alternatively, try a modified Tabata, which is normally defined as 8 rounds of HIIT with 20 seconds on and 10 seconds off, pushing your body to its limit:

- 20 seconds on (perceived 10 on the scale) and 10 seconds easy (4 or 5 on the scale). Perform continuously for 4 minutes (8 sets of 30-second intervals).
- Recover at an easy pace for 5 minutes (4 or 5 on the scale).
- Repeat 3 times.
- Cool down for 3 minutes.

Option 2: Get Glowing with HIIT (30 Minutes)

For Beginners

Do each of the following exercises. Work for 30 seconds, then rest for
10 seconds before moving to the next exercise on the list. Cycle through
each exercise three times.

For Intermediate and Advanced Levels

Increase the exercise duration and decrease the rest time between
exercises. The more advanced you are, the more you can also increase the
sets of total exercises.

High to Low Plank Holds
Get in plank position – hands and
toes on the ground, back straight, as if you were doing a push-up. Lower
your right arm so that your forearm is on the ground, then your left. Now
return to the up position, right arm first. Continue alternating arms for
the 30 seconds.

Jump Squats Stand with your feet shoulder-width apart. Squat
down, thrusting your bum backward so your thighs are about parallel
to the ground. As you return to the standing position, add a jump.
When you land, go back into the squat. Repeat for 30 seconds.

Burpees From a standing position, squat down and place your hands on the floor. Thrust your feet backward so you end up in the "up" push-up position. Do one push-up, then thrust your feet forward to the squat position. Stand and jump. Repeat. If this move is too difficult, substitute another set of High Knees (see page 270).

Jumping Lunges
Stand with one foot forward in the lunge position. Lower your body so that your front thigh is parallel to the ground. As you rise back up, jump and switch the position of your legs so the other leg is in front. Repeat. If this move is too difficult, do regular lunges, alternating legs, without the jump.

Mountain Climbers
Get into the "up" push-up position. Keeping your back straight, thrust one foot forward and then the other (like you're running in place, while keeping your hands on the ground).

High Knees Stand in place, then start
running in place, trying to get your knees
high as you pump your arms.

V-Ups Lie flat on your back. With your arms over your head, crunch
your abs to bring your straight legs and straight arms together above you.

Resistance Exercise Training (RET) Workouts

Option 1: Glow Stronger with RET (30 Minutes)

Each exercise is a compound exercise, meaning it targets multiple muscle groups. Do 3 sets of 15 repetitions each. Rest for 30 to 60 seconds between each set.

For Beginners

Practice the movement to have perfect form using only your body weight.

For Intermediate and Advanced Levels

Add weight, in the form of dumbbells or even household items like a bottle of water or a bag of rice. Start with the smallest weight and work your way up. Advanced levels can use 2- to 5 kg (5- to 10-pound) weights.

Squat to Overhead Press Stand with your feet shoulder-width apart, holding a weight in each hand. Squat so your thighs are parallel to the ground. As you stand back up, press the weights directly over your head. Return the weights to your side before you return to a squat.

Step-Up with Bicep Curl

Stand in front of a small step or bench, holding a weight in each hand. Step up with your right leg while raising your left leg as if taking a step. Curl both weights from your side (your hands should end up at shoulder level). Step back down and repeat with your left foot.

Single-Leg Bent-Over Row

Stand on one leg, hinging at the hip so you're bent over and your back is at about a 45-degree angle. Holding a weight in each hand, row the weights from a down position up to your side (your elbows will be pointing toward the ceiling). Do half the reps on one leg, half the reps on the other.

Sumo Deadlift High Pull
Stand with your feet wider than shoulder width and your toes pointed out. Hold the weights close together as your squat down, letting the weights nearly touch the floor. As you stand back up, pull the weights up toward your chin in one motion.

Shoulder Tap Push-Ups
Get in push-up position (your knees can be on the ground). Do one push-up. When you're back up, touch your right shoulder with your left hand and then your left shoulder with your right hand. That's one.

Option 2: YoGlow with RET (30 Minutes)

In this yoga-style strength training workout, aim for 3 sets of each yoga pose with weights.

For Beginners

Practice the movement to have perfect form using only your body weight.

For Intermediate and Advanced Levels

Begin this workout with 1kg (2 pound) weights and work up. Exceeding 4kg (8 pounds) is not recommended.

Mountain Pose + Lunge Start in Mountain Pose (Tadasana), holding a weight in each hand. Inhale, and as you exhale, step forward with one foot into a lunge. Make sure to keep your front bent knee over your ankle. You can bend your back knee and lift your heel. To aid in balance, step forward and slightly outside of centre. Inhale, and as you exhale, step back into Mountain pose. Repeat with the other leg. One set is 5 lunges on each side, for a total of 10 lunges.

Warrior I + Upright Row Holding Dumbbells Keep one foot flat on the floor with toes pointed out at an angle. Step your other foot out into a lunge, bending your knee. This is Warrior I. Hold a weight in each hand. Bend at your hips, keeping your back as flat as possible, and lower your chest toward your knee. Next, bend your elbows and lift the weights to your torso, making sure to pull your shoulder blades together. Lower the weights. Repeat for 5 reps, then switch legs and do 5 more on the other side.

Warrior II + Bicep Curl From Warrior I, with weights in your
hands, separate your legs further to about 1.2 metres (4 feet) apart and
rotate your back foot so it is parallel to the wall behind you. Keep your
hips, head and other foot pointed forward. Bend your front leg so your
knee is over your ankle. Next, raise and extend both arms – one in front
of you, holding the weight with your palm facing the ceiling, and one
behind you with your palm facing the floor. Inhale as you keep your
front arm stable and extended, while bending the back elbow, curling
the weight toward your shoulder. Exhale and extend both arms. Inhale
and curl your front arm while leaving your back arm extended. Exhale
and extend both arms. Inhale once more and curl both arms. Exhale and
slowly return to starting position. Repeat on the opposite side.

Warrior III + Tricep Kick-Backs Start in Warrior I, with weights in your hands. Be sure to keep your toes about 3 feet apart, with your back leg slightly angled and your front leg bent. Lower your chest toward the floor while straightening your front leg. Keep your arms as straight as possible. Your back leg should start to rise as you bend forward. Don't force it. Just lift as high as is comfortable for you. As your back leg lifts, extend your arms forward. Try to engage your core muscles to keep your back flat. Once you are stable, bring both of your arms back, in line with your torso (so your hands and feet are now all in back of your head). Now lift your arms up. This can be a very small move. Make sure to squeeze your shoulder blades together and tighten your abs. Press into your standing leg for stability. Carefully return to Warrior I. Repeat for 5 reps, then switch legs and do 5 more on the other side.

Note for beginners: You can do one arm at a time, keeping the other forward and resting on a chair for balance.

Plank Extension with Weights With a dumbbell in each hand, start in plank position, either on your knees or with legs extended, making sure your arms are parallel, your hands are shoulder-width apart and your feet are hip-width apart. Keep your head in line with your neck, looking forward, with your eyes slightly down. Tighten your core and lift into a push-up position. Extend one arm and lift it in front of you as high as you are comfortably able. Bring your arm back to plank position. Repeat for 4 reps and switch sides. For added support, keep your feet or knees slightly wider than hip-width apart.

Acknowledgments

From the very beginning stages of Glow15, I knew what I wanted to accomplish. I wanted you to live your life with boldness, with courage, with passion, with energy, and with the beauty and youthfulness that make you feel like, well, you. This project has represented not only what I have done in my career, but also what I have done throughout my life. I am honoured to bring you Glow15, and I am humbled to be surrounded by so many people who helped me do that. A book, like any project, is only as strong as all of its parts. And I wanted to thank the people who made this book come to life.

A book project like this – from conception to publication – can look like a major freeway system. So many things are travelling in different directions – sometimes at high speeds, sometimes with roadblocks.

To the beautiful and smart JJ Virgin, you introduced me to the most wonderful and amazing literary rockstar agent in Celeste. Your desire to connect the dots and people makes all the difference.

To my agent, Celeste Fine, you have been my light and vision. You got so excited about Glow15 three years ago and patiently guided me to this day.

To Stefanie Schwartz and Lindsey Green, you took my dream of creating a book and made it a reality. From the inception to production, you handled both the big picture and the details with incredible skill, creativity, and thoughtfulness.

To Ted Spiker, thank you for writing and helping to organize and guide me through the process of writing my first book. Your incredible talent around the written word is truly remarkable and it was so great to have you involved.

To Deb Brody, thank you for being bold and believing in me. Nothing made me happier than to have you decide that you would be my editor and publisher no matter what. Your vision and leadership through the process has been exactly what I was looking for. Your kindness and support—while also being fierce and decisive—have been inspiring. To the entire team at Houghton Mifflin Harcourt: Marina Padakis Lowry, Melissa Lotfy, Melissa Fisch, Tom Hyland, Rebecca Springer, Jessica Gilo, Breanne Sommer and Ivy McFadden . . . thank you for believing in me.

To Carol Brooks, my dear friend and literary mentor, your inspired and brilliant mind helped guide me. You are always willing to be generous and kind with your time and understanding of what makes us women tick. You are a beacon of light and love.

To master storyteller Mary Kinkaid, thank you for spending countless hours working with me to bring out my stories. I had no idea I had so many to share in this book.

To Sarah Passick, thank you for all the positive support and many ideas. Without you we could never have reached our goals.

To Holly Lichtenfeld, Karen Balcombe, Amy Shay-Jacobs and Jen Hansard, thank you for your dedication and hard work to get the message of Powerphenols and AutophaTea into the marketplace.

Thank you to my friends Kat Brooks, Suzanne Gatt and Rachel Simons, the three women who independently and together brought substance and love to this book through the right skincare ingredients to the exercises that make all the difference.

To the women who participated in the study and my beta test, thank you for leading us and showing us how to be brave.

One of my great joys in writing Glow15 was discovering the science behind it. I'm indebted to the masterful minds who have spent hours teaching me and all the women in the programme about autophagy. Their work is an inspiration.

Above all, I want to thank Elzbieta Janda, PhD, for introducing me to the world of autophagy. You are my personal hero, and I look forward to knowing you for the next sixty-plus (youthful) years to come.

To Nobel Prize winner Yoshinori Ohsumi, PhD, thank you for your tireless commitment to autophagy and paving the way to improving millions of people's health.

To Steve Anton, PhD, how can I thank you for all the years of tirelessly answering each and every question I have had around how women can be fit, have energy, and look and feel our absolute best? From the countless hours on weekends working from your office to going through hundreds and thousands of studies to find the perfect information, you have been my partner and dear friend.

To Heather Hausenblas, PhD, the hottest doctor on campus. Thank you for teaching us how to exercise to activate our autophagy and do so in a way that we can get past all the self-sabotage that exists. You are the greatest and a true inspiration.

In addition, many experts helped shape the content of this book, as I worked closely with them to produce scientifically backed plans and principles.

To Michael Breus, PhD, The Sleep Doctor, you took our sleep plan and worked it so that all women can harmonize their autophagy with their circadian rhythm to sleep better. How exciting to learn that your cutting-edge information is tied to the 2017 Nobel Prize in Medicine. We're deeply grateful for the information you share with us.

To William Dunn, PhD, a star in the world of science and autophagy, you guided us to understand this complex and fascinating topic.

To Dendy Engleman, MD, you provided important insight about skin and beauty. You are the true definition of beauty and brains, and you guided us with science-backed ideas to help women outsmart wrinkles.

To Richard Wang, MD, PhD, a world-renowned autophagy expert and dermatologist, through your intellect and understanding of skin and autophagy, you allowed us to put together the very best autophagy-activating ingredients and a plan that can help all women look and feel younger fast.

To clinical nutritionist Rob Maru, who is not only a natural product innovator, but also my friend – you are like a brother to me. You have the most amazing way of developing the concepts and products that change millions of people's lives. I absolutely love to share your health upgrades with women all over the world. Thank you for sharing your passion.

To Jenny Ruff – publicist and friend, style consultant, confidant and beauty guru – you worked tirelessly day or night and I knew that with you by my side, I would be able to share this book with all of your best friends.

To Tim Adams, thank you for being my coach and helping me to do the right exercises at the right times. I couldn't have done it without you.

To Jacqueline Phillips, you are the makeup artist extraordinaire. And as a woman who never likes to wear makeup, I always look forward to the way you make me look.

To Carol Perrine, thank you for helping me train my brain through the ups and downs of writing this book.

To Wayne Lukas, who's better than you? I am in constant awe of the way you can influence millions of women with your style and brilliance. Thank you from the bottom of my heart for influencing me.

To nutritionist Brooke Alpert, RD, who provided meals and recipes, you knew exactly how to take the ingredients that make us young and put them together in a delicious way.

To clinical nutritionist Lizzy Swick, MS, RDN, you always said yes and made all my dreams around food come true in simple and smart recipes.

To Michael Fishman, for bringing me into the world of online communications. Your careful curation and leadership of our community has allowed so many great voices and messages to be expressed.

To Christina Rasmussen, you are my soul sister, and the work that you do around grief helps millions of us (www.goodlifeproject.com) – thank you.

To Pedram Shojai, thank you for sharing my story of sustainability with millions of people through your documentary *Prosperity*. Keep up the good work.

To Dave Asprey, picking up the bar of Kerrygold sitting at JJ Virgin's wedding table and knowing it was yours says it all. You walk the walk and talk the talk.

To Bo and Dawn Easeman, thank you for giving me my voice and ability to communicate my story, which as you can see I started my book with and I start every conversation with.

To Dr Kellyann Petrucci, your passion, perseverance and girl-power are awesome.

To Nicole Beurkens, our families can glow and thrive only as much as we do, and your expertise in children's health and nutrition makes that possible.

To Adam Bornstein, thank you for your enthusiasm and expertise in communication. I'm forever in awe.

To Michael and Izabella Wentz, thank you for guiding me on how best to communicate my message. You two are such a powerful husband-and-wife team and through the *Thyroid Secret* and *Hashimoto's Protocol* have had a massive impact on people's health and well-being. I'm proud to know you both.

To Amber Spears, thank you for sharing my message through your world in your smart and savvy way.

To Jenny Thompson, thank you for being a CEO muse. Your mind is a fountain of truly brilliant ideas, and I am grateful to call you my friend.

To Dr Joe Maroon, a neurosurgeon, Ironman, and 77 years young, I learn so much from you. You have always been my mentor in the world of self health and well-being. I look to you to teach us about omegas, polyphenols, epigenetics and inner balance.

To Sanjay Khosla—my mentor and fellow mountain climber – every day I grow from something that you teach me about building meaningful solutions and a business with soul. You are one in a billion, and I am deeply humbled as I constantly learn from you.

To Tom Aarts, my dear friend who has opened the doors to more health and wellness. You are a force of good, and I cannot wait for the world to experience your company, Smart Family Nutrition.

To Elayna Rexrode, you have been a constant support. I love that you have adopted the Glow15 programme and made it your own. I could not have done it without you. Thank you for your nights and weekends.

Most importantly, thank you to my family, for the love, support and time that you gave me. Without it, I never would have had the opportunity to share this book with the world.

To Mum and Dad, for giving me a massive head start and showing me the principles within this book forty years ago! The lifestyle you introduced me to – that I was once embarrassed for my friends to see – is now what everyone is embracing. I am proud of that, and I'm proud to be your daughter.

To my husband, Rob, and my incredible children, Megan, Christian, Brendon and Samuel: you are my inspiration.

References

During the development of this book and programme, hundreds of studies were reviewed, and leading scientists, researchers, and experts were interviewed. Below are citations to specific works referenced in *Glow15*, as well as studies that influenced the major principles in the plan and individual chapters.

Introduction

Kantar Worldpanel, September 2015, www.kantarworldpanel.com; affiliated with Kantar Group, Data Investment Management Division of WPP.

The Nobel Prize in Physiology or Medicine 2016, Yoshinori Ohsumi, www.nobelprize.org/nobel_prizes/medicine/laureates/2016.

Hubbard BP, AP Gomes, H Dai, J Li, AW Case, T Considine, TV Riera, et al. "Evidence for a common mechanism of SIRT1 regulation by allosteric activators," *Science* 339, no. 6124 (March 8, 2013): 1216–19.

Chapter 1

Pyo JO, SM Yoo, and YK Jung. "The interplay between autophagy and aging," *Diabetes and Metabolism Journal* 37, no. 5 (October 2013): 333–39.

Remenapp A, B Broome, G Maetozo, and H Hausenblas. "Efficacy of a multiple health behavior change intervention on women's health outcomes," *Women's Health Open Journal* 3, no. 1 (2016): 8–14.

Chapter 2

Ervin RB, and CL Ogden. "Consumption of Added Sugars Among U.S. Adults, 2005–2010." US Department of Health and Human Services, Centers for Disease Control and Prevention, *NCHS Data Brief* 122

(May 2013): 1–8.

Centers for Disease Control and Prevention. National Report on Human Exposure to Environmental Chemicals, January 2017, www.cdc.gov /exposurereport/index.html.

National Sleep Foundation. Sleep Health Index 2014, https:// sleepfoundation.org/sleep-health-index-2014-highlights.

Sabatini D, and M Adesnik. "Christian de Duve: Explorer of the cell who discovered new organelles by using a centrifuge," *PNAS* 110, no. 33 (2013): 13234–35.

Madeo F, A Zimmermann, M Chiara Maiuri, and G Kroemer. "Essential role for autophagy in life span extension," *Journal of Clinical Investigation* 125, no. 1 (2015): 85–93.

Glick D, S Barth, and KF Macleod. "Autophagy: cellular and molecular mechanisms," *Journal of Pathology* 221, no. 1 (May 2010): 3–12.

Pyo J, S Yoo, and Y Jung. "The interplay between autophagy and aging," *Diabetes and Metabolism Journal* 37, no. 5 (October 2013): 333–39.

Gelino S, and M Hansen. "Autophagy – an emerging anti-aging mechanism," *Journal of Clinical and Experimental Pathology* suppl. 4 (July 12, 2012): S6.

Chapter 3

Lekli I, D Ray, S Mukherjee, N Gurusamy, MK Ahsan, B Juhasz, I Bak, et al. "Co-ordinated autophagy with resveratrol and γ-tocotrienol confers synergetic cardioprotection," *Journal of Cellular and Molecular Medicine* 14, no. 10 (October 2010): 2506–18.

Chapter 4

Catenacci VA, Z Pan, D Ostendorf, S Brannon, WS Gozansky, MP Mattson, B Martin, et al. "A randomized pilot study comparing zero-calorie alternate-day fasting to daily caloric restriction in adults with obesity," *Obesity* 24 (2016): 1874–83.

Longo VD, et al. "Fasting, circadian rhythms, and time-restricted feeding in healthy lifespan," *Cell Metabolism* 23, no. 6 (June 2016): 1048–59.

Bray MS, J Tsai, C Villegas-Montoya, BB Boland, Z Blasier, O Egbejimi, M Kueht, and ME Young. "Time-of-day-dependent dietary fat consumption influences multiple cardiometabolic syndrome parameters in mice," *International Journal of Obesity* (London) 34, no. 11 (November 2010): 1589–98.

Stracke BA, CE Rüfer, FP Weibel, A Bub, and B Watzl. "Three-year comparison of the polyphenol contents and antioxidant capacities in organically and conventionally produced apples (*Malus domestica* Bork. Cultivar 'Golden Delicious')," *Journal of Agriculture and Food Chemistry* 57, no. 11 (June 2009): 4598–605.

Padayatty SJ, HD Riordan, SM Hewitt, A Katz, LJ Hoffer, and M Levine. "Intravenously administered vitamin C as cancer therapy: three cases," *CMAJ* 174, no 7 (March 28, 2006): 937–42.

Atiya Ali M, E Poortvliet, R Strömberg, and A Yngve. "Polyamines in food: development of a food database," *Food & Nutrition Research* (January 14, 2011): 55. doi: 10.3402/fnr.v55i0.5572.

Janda E, A Lascala, C Martino, S Ragusa, S Nucera, R Walker, S Gratteri, and V Mollace. "Molecular mechanisms of lipid- and glucose-lowering activities of bergamot flavonoids," *PharmaNutrition* 4 (2016): S8–18.

Chapter 5

Kousa A, AS Havulinna, E Moltchanova, O Taskinen, M Nikkarinen, J Eriksson, and M Karvonen. "Calcium:magnesium ratio in local groundwater and incidence of acute myocardial infarction among males in rural Finland," *Environmental Health Perspectives* 114, no. 5 (May 2006): 730–34.

Davis DR, MD Epp, and HD Riordan. "Changes in USDA food composition data for 43 garden crops, 1950 to 1999," *Journal of the American College of Nutrition* 23, no. 6 (December 2004): 669–82.

Scheer R, and D Moss. "Dirt poor," EarthTalk, *Scientific American,* www.scientificamerican.com/article/soil-depletion-and-nutrition-loss.

Misner, B. "Food alone may not provide sufficient micronutrients for preventing deficiency," *Journal of the International Society of Sports Nutrition* 3 (2006):51.

Turner RS, RG Thomas, S Craft, CH van Dyck, J Mintzer, BA Reynolds, JB Brewer, et al. "A randomized, double-blind, placebo-controlled trial of resveratrol for Alzheimer disease," *Neurology* 85, no. 16 (October 20, 2015): 1383–91.

Szkudelski T, and Szkudelska K. "Anti-diabetic effects of resveratrol," *Annals of the New York Academy of Sciences* 1215 (2011): 34–39.

Gupta SC, G Kismali, and BB Aggarwal. "Curcumin, a component of turmeric: from farm to pharmacy," *Biofactors* 39, no.1 (2013): 2–13.

Sanmukhani J, V Satodia, J Trivedi, T Patel, D Tiwari, B Panchal, A Goel, and CB Tripathi. "Efficacy and safety of curcumin in major depressive disorder: a randomized controlled trial," *Phytotherapy Research* 28 (2014): 579–85.

Yin J, H Xing, and J Ye. "Efficacy of berberine in patients with type 2 diabetes mellitus," *Metabolism* 57, no. 5 (May 2008): 712–17. doi: 10.1016/j.metabol.2008.01.013.

Kong W, J Wei, P Abidi, M Lin, S Inaba, C Li, Y Wang, et al. "Berberine is a novel cholesterol-lowering drug working through a unique mechanism distinct from statins," *Nature Medicine* 10, no. 12 (December 2004): 1344–51.

Zhang Q, X Xiao, K Feng, T Wang, W Li, T Yuan, X Sun, et al. "Berberine moderates glucose and lipid metabolism through multipathway mechanism," *Evidence-Based Complementary and Alternative Medicine* (2011): 924851. doi: 10.1155/2011/924851.

Dulloo AG, C Duret, D Rohrer, L Girardier, N Mensi, M Fathi, P Chantre, and J Vandermander. "Efficacy of a green tea extract rich in catechin polyphenols and caffeine in increasing 24-h energy expenditure and fat oxidation in humans," *American Journal of Clinical Nutrition* 70, no. 6 (December 1999): 1040–45.

Naghma K, and H Mukhtar. "Cancer and metastasis: prevention and treatment by green tea," *Cancer and Metastasis Reviews* 29, no. 3 (September 2010): 435–45.

Scholey A, LA Downey, J Ciorciari, A Pipingas, K Nolidin, M Finn, M Wines, et al. "Acute neurocognitive effects of epigallocatechin gallate (EGCG)," *Appetite* 58, no. 2 (2012): 767–70.

Chapter 6

Hanlon B, MJ Larson, BW Bailey, and JD LeCheminant. "Neural response to pictures of food after exercise in normal-weight and obese women," *Medicine & Science in Sports & Exercise* 44, no. 10 (October 2012): 1864–70.

He C, R Sumpter Jr., and B Levine. "Exercise induces autophagy in peripheral tissues and in the brain," *Autophagy* 8, no. 10 (October 2012): 1548–51.

He C, MC Bassik, V Moresi, K Sun, Y Wei, Z Zou, Z An, et al. "Exercise–induced BCL2–regulated autophagy is required for muscle glucose homeostasis," *Nature* 481, no. 7382 (January 18, 2012): 511–15.

Schubert MM, S Hall, M Leveritt, G Grant, S Sabapathy, and B Desbrow. "Caffeine consumption around an exercise bout: effects on energy expenditure, energy intake, and exercise enjoyment," *Journal of Applied Physiology* 117, no. 7 (October 1, 2014): 745–54.

Robinson MM, S Dasari, AR Konopka, ML Johnson, S Manjunatha, RR Esponda, RE Carter, et al. "Enhanced protein translation underlies improved metabolic and physical adaptations to different exercise training modes in young and old humans," *Cell Metabolism* 25, no. 3 (2017): 581.

Luo L, A Lu, Y Wang, A Hong, Y Chen, J Hu, X Li, and Z Qin. "Chronic resistance training activates autophagy and reduces apoptosis of muscle cells by modulating IGF-1 and its receptors, Akt/mTOR and Akt/FOXO3a signaling in aged rats," *Experimental Gerontology* 48, no. 4 (2013): 427–36.

Madeo F, Eisenberg T, Büttner S, Ruckenstuhl C, and Kroemer G. "Spermidine: a novel autophagy inducer and longevity elixir," *Autophagy* 6, no. 1 (January 2010): 160–62.

Evans RW, and R Couch. "Orgasm and migraine," *Headache* 111, no. 6 (2001): 512–14.

Chapter 7

Horne JA, and O Ostberg. "A self-assessment questionnaire to determine morningness-eveningness in human circadian rhythms,"

International Journal of Chronobiology 4, no. 2 (1976): 97–110.

Chapter 8

Li L, X Chen, and H Gu. "The signaling involved in autophagy machinery in keratinocytes and therapeutic approaches for skin diseases," *Oncotarget* 7, no. 31 (August 2, 2016): 50682–97.

Nagar R. "Autophagy: A brief overview in perspective of dermatology," *Indian Journal of Dermatology, Venereology and Leprology* 83 (2017): 290–97.

Heffernan TP, M Kawasumi, A Blasina, K Anderes, AH Conney, and P Nghiem. "ATR–Chk1 pathway inhibition promotes apoptosis after UV treatment in primary human keratinocytes: Potential basis for the UV protective effects of caffeine," *Journal of Investigative Dermatology* 129, no. 7 (2009): 1805–15.

Baxter, RA. "Anti-aging properties of resveratrol: review and report of a potent new antioxidant skin care formulation," *Journal of Cosmetic Dermatology* 7 (2008): 2–7.

Trompezinski S, A Denis, E Schmitt, and J Viac. "Comparative effects of polyphenols from green tea (EGCG) and soybean (genistein) on VEGF and IL-8 release from normal human keratinocytes stimulated with the proinflammatory cytokine TNFα," *Archives of Dermatological Research* 295, no. 3 (July 2003): 112–116.

El-Domyati M, M Barakat, S Awad, W Medhat, H El-Fakahany, and H Farag. "Multiple microneedling sessions for minimally invasive facial rejuvenation: an objective assessment," *International Journal of Dermatology* 54, no. 12 (December 2015): 1361–69.

Other Selected References

Doulatov S, and GQ Daley. "Autophagy: it's in your blood," *Developmental Cell* 40, no. 6 (March 27, 2017): 518–20.

Jiao J, and F Demontis. "Skeletal muscle autophagy and its role in sarcopenia and organismal aging," *Current Opinion in Pharmacology* 34 (April 10, 2017): 1–6.

Harnett MM, MA Pineda, P Latré de Laté, RJ Eason, S Besteiro, W Harnett, and G Langsley. "From Christian de Duve to Yoshinori Ohsumi: More to autophagy than just dining at home," *Biomedical Journal* 40, no. 1 (February 2017): 9–22.

Anton SD, AJ Woods, T Ashizawa, D Barb, TW Buford, CS Carter, DJ Clark, et al. "Successful aging: Advancing the science of physical independence in older adults," *Ageing Research Reviews* 24, part B (November 2015): 304–27.

Petrovski G, and DK Das. "Does autophagy take a front seat in lifespan extension?" *Journal of Cellular and Molecular Medicine* 14, no. 11 (November 2010): 2543–51.

Trepanowski JF, RE Canale, KE Marshall, MM Kabir, and RJ Bloomer. "Impact of caloric and dietary restriction regimens on markers of health and longevity in humans and animals: a summary of available findings," *Nutrition Journal* 10 (October 7, 2011): 107.

De Castro, JM. "When, how much and what foods are eaten are related to total daily food intake," *British Journal of Nutrition* 102, no. 8 (October 2009): 1228–37.

Bray MS, JY Tsai, C Villegas-Montoya, BB Boland, Z Blasier, O Egbejimi, M Kueht, and ME Young. "Time-of-day-dependent dietary fat consumption influences multiple cardiometabolic syndrome parameters in mice," *International Journal of Obesity* (London) 34, no. 11 (November 2010): 1589–98.

Wegman MP, MH Guo, DM Bennion, MN Shankar, SM Chrzanowski, LA Goldberg, J Xu, et al. "Practicality of intermittent fasting in humans and its effect on oxidative stress and genes related to aging and metabolism," *Rejuvenation Research* 18, no. 2 (April 2015): 162–72.

Ma D, S Li, MM Molusky, and JD Lin. "Circadian autophagy rhythm: a link between clock and metabolism?" *Trends in Endocrinology & Metabolism* 23, no. 7 (July 2012): 319–25.

About the Author

Naomi Whittel, CNC, has travelled from spice markets in Bangalore to the Finger Lakes in New York state, from farms in Okinawa to vineyards in Bordeaux, with one main mission: to discover the most pure nutritional ingredients in the world that can help women transform their lives, so that they can thrive.

As founder of Reserveage Nutrition and CEO of Twinlab Consolidated Holdings, Whittel has unique access to world-renowned and well-known doctors and PhDs, which allows her to develop the most advanced scientifically pure and powerful products for wellness, better health longevity and natural beauty.

A leading nutritional expert, media personality and public speaker, Whittel is an internationally recognized expert in safe and efficacious, science-based, pure nutritional ingredients. She has won more than 25 industry awards for product excellence, focusing on such standards as purity, potency and identity. Named by *Prevention* magazine as the leading female innovator in the US natural products industry, she and her products have been praised by the *Wall Street Journal, Vogue, ELLE, Harper's Bazaar, ABC News*, PBS, *InStyle, Good Morning America* and *TODAY*, among many others. Whittel appears regularly on QVC with *Discover Wellness with Naomi,* and her products are found in more than 38 countries.

Whittel is also deeply engaged in social stewardship, from forging fair-trade partnerships with indigenous cultures to sponsoring future women entrepreneurs through DAWN (Developing and Advancing Women in Naturals), a not-for-profit initiative she founded in 2011. She also founded the Naomi Whittel Foundation to help underprivileged women sell their goods and products at fair-wage prices and create a cycle of growth.

Born in Switzerland, Whittel lives in Florida with her husband and their four children.

You can learn more about Naomi Whittel and subscribe to her newsletter, which includes the latest on better health, natural beauty and anti-ageing, at www.naomiwhittel.com.

Index

A

Accelerated Agers 32–9 *passim*, 50, 79, 126, 129, 130, 137, 175

acupressure 140, 142–3

acupuncture 13

acute stress 76

added sugar 33–4

adipocytes (fat cells) 131

advanced glycation end-products (AGEs) 33

afternoon slump 105

ageing 6–7, 8 (*see also* anti-ageing)

Alzheimer's disease, 33, 40, 67, 81, 87

American Academy of Dermatology 126, 130

American Aging Association 91

American Heart Association 97

amino acids 161

AMPK 85

anaerobic:
  exercise, 98
  fermentation 172
  threshold 98–9

Annals of the New York Academy of Sciences 82

anti-ageing, missing link 14–17, 40–1

antibiotics 149, 195

antibodies 102, 171

anti-inflammatories:
  Powerphenols as 76
  resveratrol as, 68–9

antioxidants 14, 67, 68, 71–2, 77, 86, 88, 129, 135, 137–8, 165, 167, 169, 172–3

anxiety 8, 22, 45
  reducing 24, 66–7, 83, 141

apoptosis 67

appetite:
  controlling 81
  gauging 108
  regulating 171
  suppressing 86, 89

Appetite 87

Archives of Dermatological Research Journal 137

arteries 33, 61, 82, 83

At-Home Self-Tests chart 158–9

atopic dermatitis 134

autophagy 6, 15–16
  described 30–42
  as cellular self-cleaner 36–8
  dysfunctional 6
  foods that activate 55–8, 60, 63–4
  glucagon triggers 57
  on/off switch for 38–9
  and plant extracts 77

Autophagy 95

AutophaTea 74, 86, 88, 95, 169, 199

B

bacteria 51, 73, 82, 128, 134, 169, 173
  beneficial 77–8, 85
  intracellular 40

baking 162–3, 164, 169

Bangalore 82

barbecuing 163

Barry's Boot Camp 98

beauty:
  bath 120
  outer 45–6
  from within 14

bedtime 108, 112, 115–18, 123, 148, 176

berberine 80, 84–6, 147, 153

bergamot 14, 65–7, 70, 72, 77

Big results, from small changes 46–9